Contents

Acknowledgements

My sincere thanks go to: my wife Christine for her assistance, support and constant encouragement; my daughter Sarah; grandchildren Matthew, Chris and Rebecca and son James and his partner Claire for their support, patience and sanity checks; my colleagues and associates past and present for their continued support of my work and motivation to continue.

Finally 'all the best for the future' to all who use this book, I trust it provides you with some of the help and motivation required to succeed in the construction industry.

Peter Brett

Introduction

National Vocational Qualifications (NVQs) in Construction

These qualifications focus on practical skills and knowledge. They have been developed and approved by people that work in the construction industry.

Construction NVQs are available in England, Wales and Northern Ireland. Scotland uses SVQs, which work in a similar way.

There are three levels of NVQs for construction crafts and operatives:

- ◆ Level 1 is seen as a 'foundation' to the construction industry, consisting of common core skills and occupational basic skills.
- ◆ Level 2 consists of common core skills and units of competence in a recognizable work role.
- ◆ Level 3 consists of further common core skills, plus a more complex set of units of competence in a recognisable work role, including some work of a supervisory nature.

Awarding body

CITB-Construction Skills and City & Guilds are the joint awarding body for the construction industry. CITB-Construction Skills are also responsible for the setting of standards for Craft and Operative NVQs.

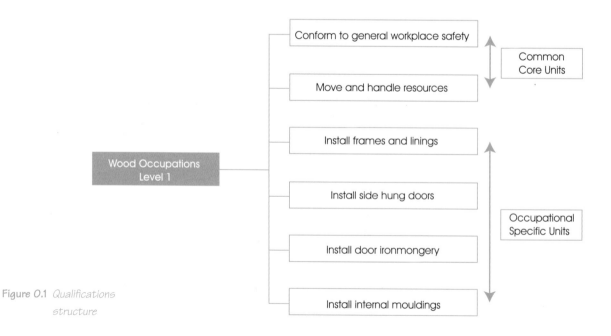

Figure 0.1 *Qualifications structure*

Unit of competence make up: in order to set out exactly what is contained in a unit and also to make it easier to assess, each unit begins with a description, for example:

Conform to General Workplace Safety

This unit is about:

- awareness of relevant current statutory requirements and official guidance;
- personal responsibilities relating to workplace safety, wearing appropriate personal protective equipment (PPE) and compliance with warning/safety signs;
- personal behaviour in the workplace;
- security in the workplace.

The description is followed by a number of statements:

Performance criteria: these state exactly what you must be able to do.

Identify hazards

Scope of performance: this sets out what evidence is required to meet each of the performance criteria. The majority of this evidence must be from the workplace, simulation evidence is only allowed in limited circumstances.

Hazards, associated with the workplace and occupations at work, are recorded and/or reported

Knowledge and understanding relating to performance criteria: this links in general terms the knowledge and understanding required to back up the performance criteria.

You must know and understand:

- the hazards associated with the occupational area;
- the method of reporting hazards in the workplace.

Scope of knowledge and understanding: this uses the key words contained in the Knowledge and Understanding statements (shown in bold type) and expands them to cover the scope of what is expected of a competent worker in the construction industry.

Hazards:

- Associated with resources, workplace, environment, substances, equipment, obstructions, storage, services and work activities.

Reporting:

- Organisational reporting procedures and statutory requirements.

Collecting evidence

You will need to collect evidence from your workplace of your satisfactory performance in each performance criteria of a unit of competence. This should be inserted in a portfolio and referenced to each unit of competence. Evidence must confirm that your practical skills meet the appropriate performance criteria. Simulation evidence in a training environment is only allowed in a limited range of topics.

Evidence can come from any of the following people:

◆ employers;   ◆ skilled work colleagues;
◆ managers;   ◆ work-based recorder;
◆ supervisors;   ◆ client.

Suitable types of evidence – you should include in your portfolio as much evidence as possible, from more than one of the following, for each performance criteria:

◆ Time sheets: detailing the work you have undertaken, for these to be valid they must be signed by yourself and the work-based recorder.
◆ Drawings: of the work you have undertaken. These should be supported by a witness testimony.
◆ Photographs: of the work you have undertaken, ideally with you in the photograph. To be valid photographs should be supported by a statement, containing a brief description of the work, details of where and when you carried it out and be signed by yourself and either the work-based recorder, manager, supervisor, skilled worker or the client.
◆ Associated documentation: used or produced as part of the work you have undertaken, such as specifications, forms and reports completed.
◆ Witness testimony: a statement by a responsible person confirming that you have undertaken certain work activities, these should include wherever possible a detailed description of the work you carried out.

Figure 0.2 *Evidence of 'me' wearing PPE*

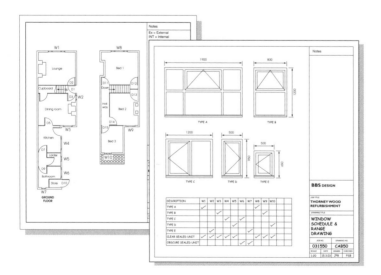

Figure 0.3 *Extracts from a specification*

Figure 0.4 *Building schedule*

> **T. Joycee Construction**
> Ridge House
> Norton road
> Cheltenham
> GL 59 1DB
>
> To whom it may concern:
>
> I can confirm, that between 15 March and 8 November 2006 James Oakley worked on the refurbishment contract at The Rivermead Estate.
>
> James was involved in the replacement of casement windows and internal window boards. He carried this work out to a competent standard at all times.
>
> This work was undertaken in occupied houses, feedback from the tenants concerning James's communication with them and his consideration shown to their property, including the cleanliness of his work was always exemplary.
>
> In addition James assisted me in the general day-to-day organization of the working environment, including the scheduling of the work and the safety induction of new staff. Indeed he always set a fine example by wearing at all times his safety helmet, boots and high visibility vest.
>
> Yours Faithfully
>
> *Chris Heath*
> Chris Heath
> (Site Project Manager)

Figure 0.5 *Witness testimony*

Where an assessor considers your evidence as insufficient in either quality or quantity, you may be asked to undertake simulated activities in order to demonstrate/reinforce your competence in particular performance criteria.

The assessment process

The joint awarding body CITB-Construction Skills and City & Guilds approves organisations to carry out assessment of people for an NVQ award in construction. Typically these are:

- further education colleges;
- private training providers;
- construction companies.

Once approved these are known as assessment organisations. Their assessment work will involve the following personnel:

- **Assessors** – these are people who are occupationally competent in the work role in which you are being assessed and also qualified in the assessment process. Their role is to decide whether you are competent in each performance criterion. They will also observe you in the workplace to ensure you are carrying out the full range of activities to create required evidence portfolio.
- **Internal verifier** – this is the person who is responsible, in an assessment organisation, for ensuring the quality of the assessments carried out by the assessors.
- **Work-based recorders** – these are people in the workplace whose employer has given them the responsibility of authenticating the evidence that a candidate is collecting for a portfolio.
- **External verifiers** – they are employed by the joint awarding body to monitor the whole assessment process and ensure that each assessment organisation is working to the standards set.

How to use this book

This book covers the four occupational specific skill units for Wood Occupations at Level 1. In addition, introductory chapters have been included on: Materials, Hand tools, Basic joints and Portable powered hand tools. Although the content of these chapters is not assessed directly as separate units of competence, knowledge of their contents is assumed and assessed in other units, at both Levels 1 & 2.

These books are intended to be supported by:

- classroom activities;
- tutor reinforcement and guidance;
- group discussion;
- films, slides and videos;
- text books;
- independent study/research;
- practical activities.

You will be working towards one or more units at a time as required. Discuss each unit's content with your group, tutor, or friends wherever possible. Attempt to answer the learning activities for that unit. Progressively work through all the units, discussing them and answering the assessment activities as you go. At the same time you should be working on the matching practical activities in the workplace and collecting the required evidence.

This process is intended to aid learning and enable you to evaluate your understanding of the particular topic and to check your progress through the units. Where you are unable to answer a question, further reading and discussion of the topic are required.

Independent study/research

'Browsing the Internet' via a computer is an excellent means of accessing other sources of information as part of your research: simply type in the website address of the company or organisation into a web browser and you will be connected to their website.

Try some of the following sites:

- Building Regulations: www.planningportal.gov.uk;
- British Standards: www.bsi-global.com;
- Building Research Establishment: ww.bre.co.uk;
- construction training and careers: www.citb-constructionskills.co.uk and www.city-guilds.co.uk;
- Government publications: www.tso.co.uk;
- health and safety: www.hse.gov.uk;
- building materials and components: www.buildingcentre.co.uk;
- types and use of timber in construction: www.trada.co.uk;
- employment rights and trade unions: www.worksmart.org.uk.

If you don't know the exact website address of the organisation you are looking for, or you simply wish to find out more information on a subject, you could use a 'search engine' to find the web pages. Search engines use 'key words' to find information on a subject. Enter a key word or words such as doors, windows, stairs or strength grading or timber etc. or the name of a company/organisation, and it searches the Internet for information about your key words or name. You are then presented with a list of relevant websites that you can click on, which link you to the appropriate information pages.

Types of learning activity

The learning activities used in this book should be completed on loose-leaf paper and included as part of your portfolio of evidence. They are divided into the following:

◆ Measuring up. Questions at the end of a major topic or units, which enable you to evaluate your understanding of a recently completed topic and to check your progress through the units. 'Measuring-up' questions are either multiple-choice questions or short answer questions.
◆ Activity. An extended learning task normally at the end of a unit, which has been designed to reinforce your technical and communication skills in day-to-day work situations.

Multiple-choice questions normally consist of a statement or question followed by four possible answers. Only one answer is correct; the others are distracters. You have to select the most appropriate letter as your response.

Example 1:

The depth of a housing joint should be restricted to about:

a) Half the timber thickness
b) One third of the timber thickness
c) A quarter of the timber thickness
d) Three quarters of the timber thickness

As one third of the timber thickness is the correct answer, your response should be (b).

Example 2:

The head of screw illustrated is:

a) Countersunk
b) Round
c) Raised
d) Bugle.

The correct answer is again (b).

Occasionally variants on the four-option multiple-choice question are used, as in the following examples.

Example 3:

Match the items in *list one* with the items in *list two*.

List one refer to illustration:

List two:

(1) adjusting nut
(2) cutting iron
(3) back iron
(4) lever cap
(5) lateral adjusting lever
(6) blade tensioner

The correct match is

	V	W	X	Y	Z
a)	4	2	3	5	6
b)	6	2	3	1	5
c)	2	3	1	5	6
d)	4	3	2	5	1

This question requires you to work through the lists matching the items up (it is usual for the lists to be of different lengths). In this example:

V is 4
W is 3
X is 2
Y is 5
Z is 1

Therefore the correct response is (d).

Example 4:

Statement: The point of a nail should be blunted when nailing near the end of a piece of timber.

Reason: The blunted point tears its way through the timber preventing it from splitting.

a) statement true   reason true
b) statement false   reason false
c) statement true   reason false
d) statement false   reason true

This type of question comprises a statement followed by a reason, where the statement can be true or false and the reason can be true or false. You are required to select the appropriate response.

In this example both the statement and reason are true, therefore the correct response is (a).

measuring up

Short-answer questions consist of a task to which a short written answer is required. The length will vary depending on the 'doing' word in the task:

◆ 'name' or 'list' normally require one or two words for each item;
◆ 'state', 'define', 'describe' or 'explain' will require a short sentence or two;
◆ 'draw' or 'sketch' will require you to produce an illustration.

In addition, sketches can be added to any written answer to aid clarification.

Example 5:

Name the timber trim that is used to cover the intersection between the wall and floor.

Typical answer: skirting.

Example 6:

Define the term 'arris' when applied to a timber section and state why it should be removed from the edges of a door.

Typical answer: An arris is the sharp external corner of a timber section, where the face and edge join. It should be removed to soften the corners and provide a better surface for the subsequent paint or lacquered finish.

Example 7:

Produce a sketch to show the difference between laminboard and blockboard.

Typical answer:

Laminboard

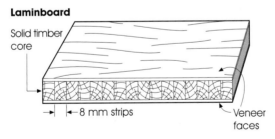

Solid timber core

← 8 mm strips →

Veneer faces

Blockboard

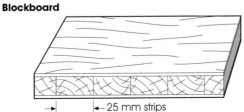

← 25 mm strips →

How to use this book

activity

These activities are a combination of short-answer questions on the same topic. They normally commence with a statement containing a certain amount of background information designed to set the scene for the question. This is then followed by a series of questions in logical order. The length of the expected answer to each sub-part will vary, depending on the topic and the wording of the question, from one or two words to a paragraph. There may be a blank form to complete, a sketch, a calculation, or a combination of any of these. At each stage the wording of the question will make it clear what is required. Blank forms for completion and inclusion in your portfolio of evidence may be downloaded from www.nelsonthornes.com/carpentry

Example 8:

After felling, timber is converted into usable sizes and then seasoned ready for use.

(a) Produce large end grain sketches to show:
  (i) A radial sawn plank
  (ii) A tangential sawn plank
(b) Show the shape that each plank is likely to take up as a result of shrinkage.
(c) State what is meant by the term 'equilibrium moisture content'.
(d) Name *two* methods of determining the moisture content of a plank.
(e) State *four* reasons for seasoning timber.
(f) Name *two* distortions that can occur during seasoning.
(g) State the probable causes of the defect.

Typical Answer
(a) (i) Radial cut. Annual rings at 45° or more
     (ii) Tangential cut. Annual rings at less that 45°

(b) (i)

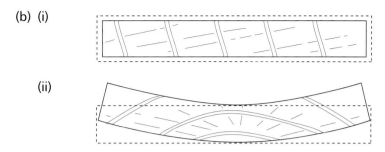

(ii)

(c) When a piece of timber has a moisture content that is equal to the surrounding atmosphere, it is said to have an equilibrium moisture content.

(d) (i) Oven-drying method
    (ii) Electric moisture meter

(e) (i) To ensure the moisture content of the timber is below the dry rot safety line of 20 per cent.
    (ii) To ensure that any shrinkage takes place before the timber is used.
    (iii) Using seasoned timber, the finished article will be more reliable and less likely to split or distort.
    (iv) Wet timber will not readily accept glue, paint or polish.

(f) (i) Bow
    (ii) Twist

(g) Distortions can be caused as a result of poor stacking or bad air circulation.

In addition to completing the learning activities, you may be asked oral questions by your tutor, assessor or verifier. This is often done to gain further evidence of your written response or questions may be asked during a review of your portfolio to gain supplementary evidence: these questions normally take the form of 'How did you… ?' 'Why did you… ? 'What would you do in the following circumstances… ? etc.

Other learning features used in this book

These include the following:

Colour enhanced illustrations and documents as an aid to clarity and reinforcement of text.

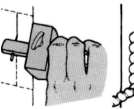

Mark position on door edge   Gauge centre line on door edge   Drill out to width and depth

did you know?

Units of competence
– contain the work-related skills and knowledge needed to do a job effectively.

Assessment – is the process of measuring your level of knowledge, understanding and work-based performance to show that you are competent in an area of your work.

Always ensure that a power tool is disconnected from the power supply before making any adjustments to guards, before changing any tooling, and before undertaking maintenance.

Did you know boxes in the margin, which define new words or highlight key facts.

Safety Tip boxes in the margin, highlighting facts for you to follow or be aware of when undertaking practical tasks.

Example

Worked examples included in the text for use as a guide when answering questions or undertaking other tasks.

Determine the total length of skirting required for the room illustrated in Figure 8.51.

Perimeter = 2 + 3.6 + 2.5 + 1.6 + 0.5 + 2 = 12.2 m

Total metres run required = 12.2 – 0.8 (door opening) = 11.4 m

An allowance of 10% for cutting and waste is normally included in any estimate for horizontal moulding.

Determine the total metres run of skirting required for the run shown including an allowance of 10% for cutting and waste.

Total metres run required = 11.4 m

Total metres run required including a 10% cutting and waste allowance = 11.4 × 1.1 = 12.54 m

say 12.5 m

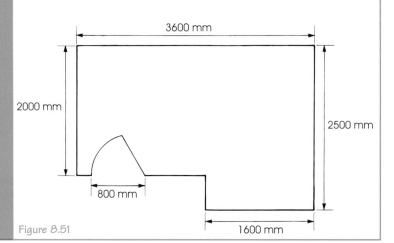

Figure 8.51

Materials

1

This chapter is intended to provide the reader with an overview of the types of timber and wood-based board material used by the woodworker. Although its content is not assessed directly, knowledge of its contents is assumed and assessed in Wood Occupations, units VR 05, VR 06, VR 07, and VR 08 at Level 1. Knowledge is also assessed in both Site Carpentry and Bench Joinery NVQ units at Level 2.

In this chapter you will cover the following range of topics:

- How a tree grows
- Tree types
- The structure of wood
- Converting wood
- Drying wood
- Wood defects
- Preservation
- Common woods and their main uses
- Purchasing wood
- Grading of wood
- Wood-based board materials
- Woodscrews
- Nails
- Woodworking adhesives

What's required in VR 05, VR 06, VR 07 & VR 08?

To successfully complete these units you will be required to demonstrate the appropriate skill and knowledge necessary to:

- State the characteristics, uses and limitations of materials used by the woodworker.

You will be required practically to:

- Identify softwoods, hardwoods, wood-based board materials, fixings and adhesives.
- Identify defects in materials.
- Work materials with both hand and power tools.
- Calculate quantities of material.

There are many different types of wood and board materials available, each having their own specific qualities, properties and uses.

In many circumstances you will be told what particular type of wood or board to use by the client. You will have to identify the required material and select from stock the most suitable pieces, with regards to defects.

Alternatively, you may be asked to recommend a suitable material for use in a given situation. This requires knowledge of different wood properties, how it has been sawn, stored or manufactured and the nature or presence of any defects.

How a tree grows

Trees grow by adding a new layer of cells below the bark each growing season. In temperate climates, such as Europe, this is mainly during the spring and summer months, whereas in tropical climates, growth may be almost continuous or be linked to their dry and rainy seasons.

Tree parts

All trees have three common parts (Figure 1.1).

Roots – Absorb groundwater containing diluted minerals (sap) from the soil, via their system of fine hair-like ends. In addition roots act as an anchor to secure the tree in the ground. The depth and spread of the root system will depend on the type of tree and ground conditions. The radius spread of the roots will often exceed that of the crown.

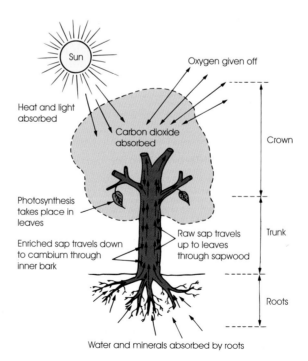

Figure 1.1 *The growth of a tree*

Trunk – The main stem of the tree, from which timber is cut. It conducts raw sap up from the roots to the crown, through the sapwood. Once modified or enriched in the leaves it travels back down again through the inner bark to the cambium layer.

Crown – The branches and leaves of a tree, often termed the canopy. It is here that the unenriched sap is processed into food for growth. This processing is called photosynthesis. The green substance in each leaf, called chlorophyll, absorbs daylight energy to convert a mixture of carbon dioxide from the air and unenriched sap from the roots into carbohydrates (sugars and starches). During this process oxygen is given off as a waste product.

Tree trunk cross-section

An understanding of the various parts is desirable as they can affect the way in which timber is cut and used (Figure 1.2).

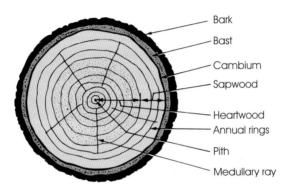

Bark
Bast
Cambium
Sapwood
Heartwood
Annual rings
Pith
Medullary ray

Figure 1.2 *Cross-section of a tree trunk*

Bark – The often corky outer layer of the whole tree. It serves to protect the inner parts from the elements, mechanical abrasion, fungal and insect attack.

Bast – Also termed inner bark. It transports the enriched sap down from the crown to all the growing parts of the tree. The layers of bast progressively become inactive as a food carrier and form a new layer of bark, with the existing bark scaling off as the tree increases in girth (circumference of the trunk) and height.

Cambium – The thin layer of living cells under the bast. These cells divide during the growing season to form both new sapwood cells and new bast cells.

Sapwood – The newly formed outer layers of growth which convey the rising, unenriched sap up to the crown. As the sapwood contains foodstuffs, timber cut from it is considered prone to insect and fungi attack, unless treated with a preservative.

Heartwood – The inner, more mature part of the trunk that no longer conveys the sap. Its main function is to give strength to the tree. Timber cut from the heartwood is often darker in colour and more durable than sapwood.

Materials Chapter 1

Pith – Also called medulla. It is the first growth of the tree as 'a sapling' and often decays as the tree ages, creating a defect in cut timbers.

Rays – Also called medullary rays as they appear to radiate from the pith outwards towards the bark. They are groups of food storage cells. Some timbers are specially cut to show the rays on the planed surface for a decorative effect.

Growth rings

did you know?

You can count growth rings to find out how long a tree has been growing.

Growth rings are also termed annual rings. In temperate climates such as Europe, tree growth takes place mainly during the spring and summer months. This produces a ring of new growth each year. By counting the number of rings you can tell how long the tree has been growing.

In tropical countries a tree's growth can be almost continuous or linked to other climatic conditions. On some occasions it is virtually impossible to distinguish between each year's growth.

In trees where growth rings are visible, it is normally possible to see that each ring in fact consists of two rings (Figure 1.3): early wood (spring growth) with thin-walled cells for maximum food transfer and appearing lighter in colour; late wood (summer growth) with thicker walled cells, as growth is slowing, which appear darker.

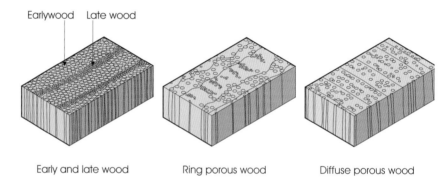

Earlywood   Late wood

Early and late wood   Ring porous wood   Diffuse porous wood

Figure 1.3 *Growth Rings*

Tree types

Trees can generally be divided into two groups:

◆ Softwood trees have narrow or needle leaves (conifers) (Figure 1.4).
◆ Hardwood trees have broad leaves (broadleaf) (Figure 1.5).

did you know?

The simple view that all hardwoods are hard and all softwoods are soft, cannot be relied on.

This grouping has little to do with the hardness of the timber but is based on the trees cellular structure. However most hardwoods are hard and heavy; likewise most softwoods are less hard and lighter in weight but these terms can be misleading as there are exceptions. Balsa wood, classified as a hardwood, is much softer and lighter in weight than say pitch pine or larch, which are classified as softwoods, which in turn are much harder and heavier than many hardwoods. Generally, only a clear difference between softwood and hardwood trees can be seen in the standing tree.

Figure 1.4 *General characteristics of softwood trees*

Figure 1.5 *General characteristics of hardwood trees*

Tree shapes

The trunks of forest or dense woodland trees will be longer and relatively free of side branches, as the crown attempts to reach up above other adjacent trees in competition for light (Figure 1.6).

General identifying characteristics of softwood trees

did you know?

Trees from dense woodlands will have a longer, straighter trunk.

- ◆ A trunk, which is very straight and cylindrical with an even taper.
- ◆ A crown that is narrow and pointed.
- ◆ Needle-like leaves.
- ◆ A bark, which is coarse and thick.
- ◆ The seeds are in cones.
- ◆ They are evergreen (they do not drop all their leaves at once in the autumn).

Materials Chapter 1

Forest grown hardwood   Forest grown softwood

Figure 1.6 *Tree shapes*

General identifying characteristics of hardwood trees

- ◆ An irregular, less cylindrical trunk, which very often has little taper.
- ◆ A crown, which is wide, rounded and contains large heavy branches.
- ◆ Broad leaves.
- ◆ The bark varies widely. It can be very smooth and thin to very coarse and thick, and range from white to black in colour.
- ◆ Covered seeds, e.g. berries, acorns and stoned fruits.
- ◆ Mainly deciduous (they shed their leaves in winter).

The structure of wood

Apart from the differences in general identifying characteristics, there are more important differences between the cellular structure of the two types of tree.

Hardwood structure

Hardwood trees have a more complex structure than softwood consisting of three types of cell: fibres, parenchyma and vessels or pores (Figure 1.7).

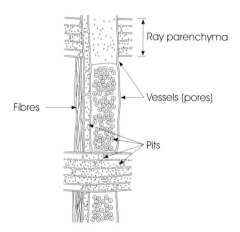

Figure 1.7 *Vertical section of hardwood*

Fibres – the main structural tissue of the wood, giving it its mechanical strength.

Parenchyma cells – the food storing cells which radiate from the centre of the tree. These are also known as ray parenchyma or medullary rays.

Vessels or pores – sap-conducting cells. These cells are circular or oval in section and can occur in the annual ring in one of two ways:

- They can be of a uniform size and be spread or diffused fairly evenly throughout the year's growth. Trees of this type are known as diffuse porous hardwoods.
- They can form a definite ring of large cells in the early spring growth, with smaller cells spread throughout the growing season. These are known as ring porous hardwoods.

Fast grown, ring porous hardwoods are considered to be stronger than slow grown, ring porous hardwoods (Figure 1.8). This is because there is less room for the strength-giving fibres between the pores in the slow grown timber.

Softwood structure

Softwood trees have a simple structure, with only two types of cell: tracheids and parenchyma (Figure 1.9).

Tracheids – box-like cells, which form the main structural tissue of the wood and, as well as giving the tree its mechanical strength, they also conduct the rising sap. The sap passes from one tracheid to another through thin areas of the cell's wall known as pits. Tracheids are formed throughout the tree's growing season, but the rapid spring growth produces a wide band of thin-walled cells (called early wood). These cells conduct sap but provide little strength. It is the summer growth of thick-walled cells (called late wood) that provides most of the tree's mechanical strength.

Parenchyma – cells that perform the same function as they do in hardwood, i.e. storing food.

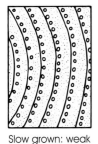

Slow grown: weak

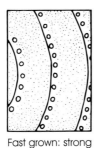

Fast grown: strong

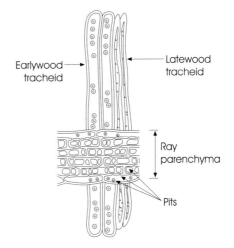

Earlywood tracheid

Latewood tracheid

Ray parenchyma

Pits

Figure 1.8 *Fast and slow grown hardwood*

Figure 1.9 *Vertical section of softwood*

Materials

Chapter 1

Resin ducts or pockets are also found in softwood but perform no useful function.

Slow grown softwoods are considered to be stronger than fast grown softwoods. This is because the slow grown timber contains more thick-walled, strength-giving tracheids than the fast grown (Figure 1.10).

Slow grown: strong   Fast grown: weak

Figure 1.10 *Fast and slow grown softwood*

Converting wood

The conversion of wood is the sawing up or breaking down of the tree trunk into various sized pieces of timber for a specific purpose (Figure 1.11).

Theoretically a tree trunk can be sawn to the required size in one of two ways, by sawing in a:

- tangential direction
- radial direction.

The terms 'radial' and 'tangential' refer to the cut surfaces in relation to the growth rings of the tree. Both methods have their own advantages and disadvantages.

In practice very little timber is cut either truly tangentially or radially because there would be too much wastage in both timber and manpower.

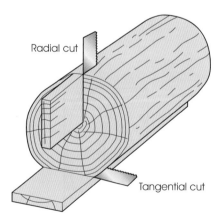

Radial cut

Tangential cut

Figure 1.11 *Timber conversion*

Timber conversion and sawing standards

To be classified as either a tangential or a radial cut the timber must conform to the following standards.

Tangential wood

Converted so that the annual rings meet the wider surface of the wood over at least half its width, at an angle of less than 45° (Figure 1.12).

Radial wood

Converted so that the annual rings meet the wider surface of the wood throughout its width at an angle of 45° or more (Figure 1.13).

There are four main methods of conversion to produce wood to these standards:

- through and through
- tangential
- quarter
- boxed heart.

Through and through – also known as slash or slab sawing (Figure 1.14). This is the simplest and cheapest way to convert wood, with very little wastage. Approximately two-thirds of the boards will be tangential and one-third (the middle boards) will be radial. The majority of boards produced in this way are prone to a large amount of shrinkage and distortion.

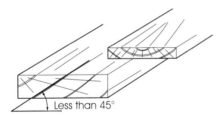

Less than 45°

Figure 1.12 *Tangential cut*

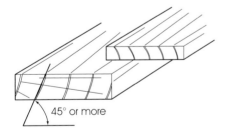

45° or more

Figure 1.13 *Radial cut*

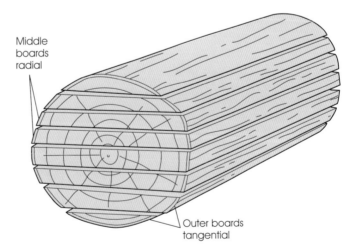

Middle boards radial

Outer boards tangential

Figure 1.14 *Through and through (slab sewn)*

Tangential – used when converting wood for floor joists and beams, since it produces the strongest timber (Figure 1.15). It is also used for decorative purposes on woods, which have distinctive annual rings, e.g. pitch pine and Douglas fir, because it produces 'flame figuring' or 'fiery grain' (Figure 1.16).

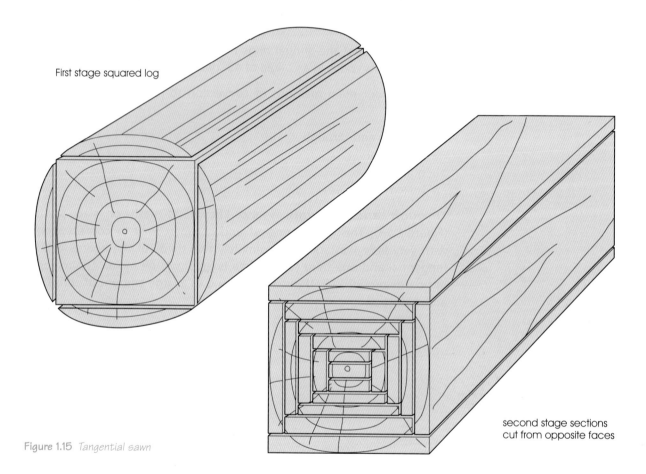

First stage squared log

second stage sections cut from opposite faces

Figure 1.15 *Tangential sawn*

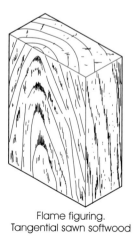

Flame figuring.
Tangential sawn softwood

Figure 1.16 *Flame figuring*

Quarter – the most expensive method of conversion, although it produces the best quality wood which is ideal for joinery purposes (Figure 1.17). This is because the boards have very little tendency to shrink or distort. In timber where the medullary rays are prominent, such as oak, the boards will have a silver figured finish; others with interlocking grain, such as African mahogany, will show a stripe or ribbon grain (Figure 1.18).

Boxed heart – when the heart of a tree is rotten or badly shaken it may be boxed in to keep the defect within the central section. Either radial or tangential boards may be produced.

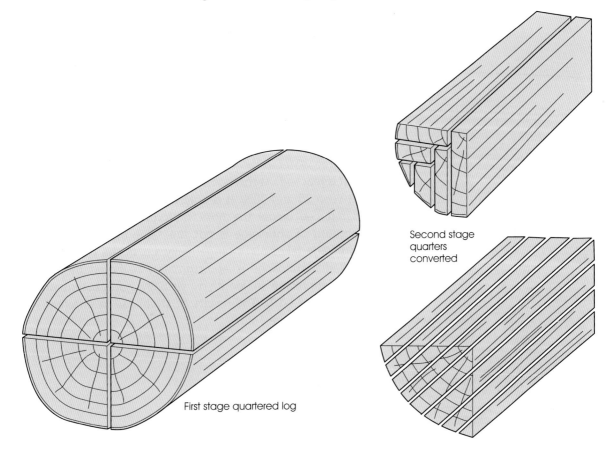

Second stage quarters converted

First stage quartered log

Figure 1.17 *Quarter sawn*

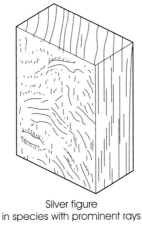
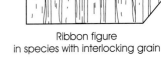

Silver figure
in species with prominent rays

Ribbon figure
in species with interlocking grain

Figure 1.18 *Decorative figure in quarter sawn timber*

Drying wood

did you know?

Distortion on drying is dependent on the orientation of the growth rings.

Seasoning – refers to the controlled drying of timber by natural or artificial means of converted sections. There are many reasons why seasoning is necessary, the main ones being:

◆ To ensure the moisture content of the timber is below the dry rot safety level of 20%.
◆ To ensure that any shrinkage takes places before the wood is used.
◆ Dry wood is easier to work with than wet wood.
◆ Using seasoned wood, the finished article will be more reliable and less likely to split or distort.
◆ In general, dry wood is stronger and stiffer than wet wood.
◆ Wet wood will not readily accept glue, paint or polish.

Moisture – this occurs in wood in two forms:

◆ as free water in the cell cavities;
◆ as bound water in the cell walls.

When all of the free water in the cell cavities has been removed, the fibre saturation point is reached. At this point the wood normally has a moisture content of between 25 and 30%. It is only when the moisture content of the timber is reduced below the fibre saturation point that shrinkage occurs. The amount of shrinkage is not the same in all directions. The majority of shrinkage takes place tangentially, i.e. in the direction of the annual rings. Radial shrinkage is approximately half that of tangential shrinkage, while shrinkage in length is virtually non-existent and can be disregarded (Figure 1.19). The results of shrinkage in different wood sections are shown in Figure 1.20.

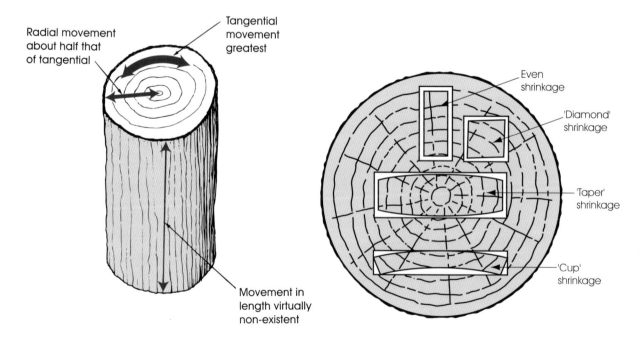

Figure 1.19 *Shrinkage*

Figure 1.20 *Results of shrinkage*

The wood should be dried out to a moisture content that is approximately equal to the surrounding atmosphere in which it will be used. This moisture content is known as the equilibrium moisture content and, providing the moisture content and temperature of the air remain constant, the wood will remain stable and not shrink or expand. But in most situations the moisture content of the atmosphere will vary to some extent and sometimes this variation can be quite considerable (Figure 1.21).

Wood fixed in a moist atmosphere will absorb moisture and expand. If it is then fixed in a dry atmosphere the bound moisture in the cells of the wood will dry out and the timber will start to shrink. This is exactly what happens during seasonal changes in the weather. Therefore all wood is subject to a certain amount of moisture movement and this must be allowed for in all construction and joinery work.

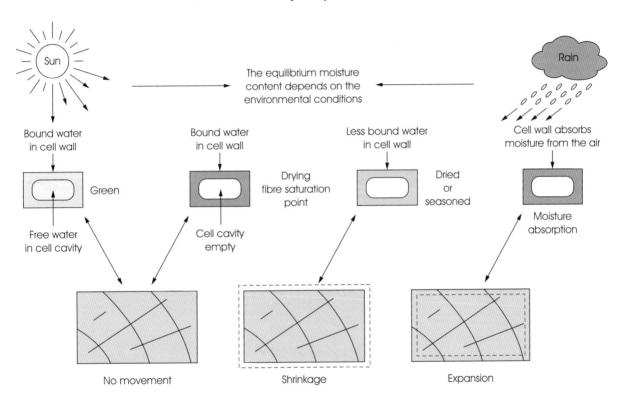

Figure 1.21 *Fibre saturation and moisture*

The moisture content of wood is expressed as percentage. This refers to the weight of the water in the wood compared to the dry weight of the wood (Figures 1.22 and 1.23).

In order to determine the average moisture content of a stack of wood, select a board from the centre of the stack, cut the end 300 mm off and discard it as this will normally be dryer than sections nearer the centre. Cut off a further 25-mm sample and immediately weigh it. This is the wet weight of the sample. Place this sample in a small drying oven and remove it periodically to check its weight. When no further loss of weight is recorded, assume this to be the dry weight of the sample.

Materials

Chapter 1

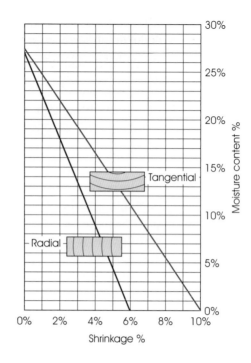

Figure 1.22 *Shrinkage compared to moisture content*

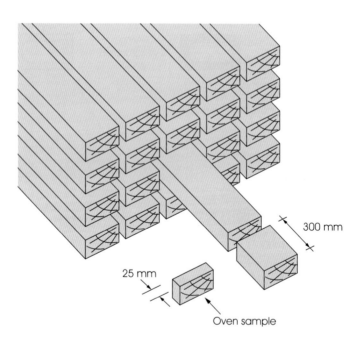

300 mm

25 mm

Oven sample

1. Weigh sample
= wet weight

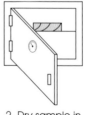

2. Dry sample in
oven until no further
loss of weight

3. Re-weigh sample
= dry weight

Figure 1.23 *Measuring moisture content of timber*

The moisture content of a piece of wood can now be found by using the following formula:

$$M/C = \frac{\text{Wet weight} - \text{Dry weight}}{\text{Dry weight}} \times 100$$

example

Wet weight of sample 50 g

Dry weight of sample 40 g

Moisture content $= \dfrac{50 - 40}{40} \times 100 = 25\%$

An alternative way of finding the moisture content of wood is to use an electric moisture meter (Figure 1.24). Although not as accurate, it has the advantage of giving an on-the-spot reading and it can also be used for determining the moisture content of wood already fixed in position. The moisture meter measures the electrical resistance between the two points of a twin electrode, which is pushed into the surface of the wood. Its moisture content can then easily be read off a calibrated dial or LED display for practical decisions to be made concerning its use (Figure 1.25).

Figure 1.24 *Moisture meter*

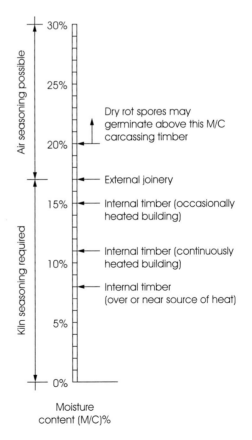

Figure 1.25 *Moisture content for various locations*

Seasoning

Wood may be seasoned in one of two ways:

- by natural means (air seasoning);
- by artificial means (kiln seasoning).

Air seasoning

In this method the wood is stacked in open-sided, covered sheds which protect the wood from rain whilst still allowing a free circulation of air. In Britain a moisture content of between 18% and 20% can be achieved in a period of two to twelve months, depending on the size of the wood.

A typical stack for air seasoning (Figure 1.26) should conform to the following:

- Brick piers and timber joists keep the bottom of the stack well clear of the ground and ensure good air circulation underneath.
- The boards are laid horizontally, largest at the bottom, smallest at the top, one piece above the other. This reduces the risk of timber distorting as it dries out.
- The boards on each layer are spaced approximately 25 mm apart.
- Piling sticks or stickers are introduced between each layer of timber at approximately 600 mm distances, to support the boards and allow a free air circulation around them. The piling sticks should be the same type of wood as that being seasoned, otherwise staining may occur.
- The ends of the boards should be painted or covered with strips of wood to prevent them from drying out too quickly and splitting.
- Hardwood can be seasoned in the same air-seasoning sheds but the boards should be stacked in the same order as they were cut from the log (Figure 1.27).

Kiln seasoning

The majority of wood for internal use is kiln seasoned, as this method, if carried out correctly, is able to safely reduce the moisture content of the wood to any required level, without any danger of degrading (causing defects). Although wood can be completely kiln seasoned, sometimes

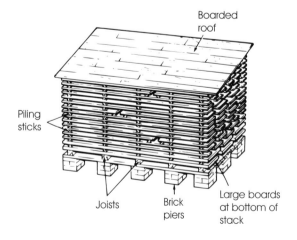

Figure 1.26 *Timber stack or shed*

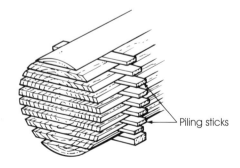

Figure 1.27 *Hardwood stacked for seasoning*

when a sawmill has a low kiln capacity, the timber is air seasoned before being placed in the kiln for final seasoning. The length of time the timber needs to stay in the kiln normally varies between two days and six weeks according to the type and size of wood.

There are two main types of kiln in general use:

◆ compartment kiln;
◆ progressive kiln.

The compartment kiln – is normally a brick or concrete building in which timber is stacked (Figure 1.28). The timber will remain stationary during the drying process, while the conditions of the air are adjusted to the correct levels as the drying progresses. The timber should be stacked in the same way as that used for air seasoning.

The drying of wood in a compartment kiln depends on three factors:

◆ Air circulation, which is supplied by fans.
◆ Heat, which is normally supplied by heating coils through which steam flows.
◆ Humidity (controlled moisture content of the air). Steam sprays are used for maintaining the humidity. They are installed along the whole length of the compartment to achieve slow seasoning.

The progressive kiln – can be thought of as a tunnel, full of open trucks containing wood which is progressively moved forward from the loading end to the discharge end. The drying conditions in the kiln become progressively more severe so that loads at different distances from the loading end are at different stages of drying.

Progressive kilns are mainly used in situations where there is a need for a continuous supply of wood, which is of the same species and dimensions.

Drying schedules

These are available for the kiln drying of different types of timbers. They set out the drying conditions required for a given size and type of timber. Although all types of timber require different conditions for varying lengths of time, the drying process in general involves three stages, these being:

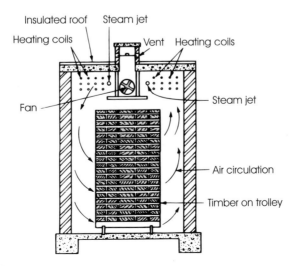

Figure 1.28 *Compartment kiln*

- Wet timber inserted – controls set to high steam, low heat.
- Timber drying – controls set to reduce steam, increase heat.
- Timber almost dry – controls set to low steam, high heat. The seasoned timber can then be removed from the kiln.

Second seasoning

This is rarely carried out nowadays, but refers to a further drying of high-class joinery work after it has been machined and loosely framed up, but not glued or wedged. The framed joinery is stacked in a store, which has a similar moisture content to the building where it will finally be fixed. Should any defects occur during this second seasoning, which can last up to three months, the defective component can easily be replaced.

Water seasoning

This is not seasoning as we understand it at all. It refers to logs, which are kept under water before conversion in order to protect them from decay. This process is also sometimes used to wash out the sap of some hardwoods, which is particularly susceptible to attack by the lyctus beetle.

Conditioning

This is a form of reverse seasoning. It refers to the practice of brushing up to one litre of water on the back of hardboard 24 to 48 hours before fixing. This is so that the board will tighten on its fixings as it dries out and shrinks. If this were not done expansion could take place, which would result in the board bowing or buckling.

Storage of seasoned timber

As the seasoning of wood is a reversible process, great care must be taken in the storage of seasoned timber. Carcassing wood and external joinery will normally be delivered to a site at an early stage. This should be stacked clear of the ground using piling sticks between each layer or item and covered with waterproof tarpaulins (Figure 1.29). Internal joinery items should not be delivered to a site until a suitable store is available (Figure 1.30). Before low moisture timber is delivered or installed, the

did you know?

Seasoning is a reversible process. Always store timber in the condition that it will be used.

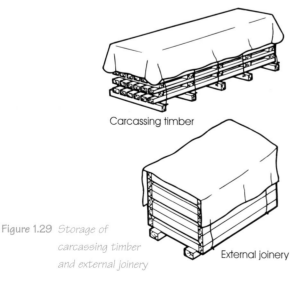

Carcassing timber

External joinery

Figure 1.29 *Storage of carcassing timber and external joinery*

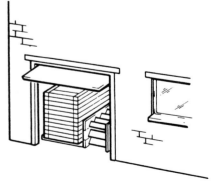

Figure 1.30 *Storage of internal joinery*

Chapter 1 Materials

building should be fully glazed and its heating system in operation, or if this is not possible a temporary heating system should be used to dry out the building and maintain a low humidity.

Wood defects

Wood is subject to many defects, which should, as far as possible, be cut out during its conversion. These defects can be divided into two groups:

◆ seasoning defects;
◆ natural defects.

Seasoning defects

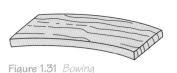

Figure 1.31 *Bowing*

Bowing – is a curvature along the face of a board, and often occurs where insufficient piling sticks are used during seasoning (Figure 1.31).

Figure 1.32 *Springing*

Springing – is a curvature along the edge of the board where the face remains flat. It is often caused through bad conversion or curved grain (Figure 1.32).

Winding – is a twisting of the board and often occurs in wood that is not converted parallel to the pith of the tree (Figure 1.33).

Cupping – is a curvature across the width of the board and is due to the fact that wood shrinks more tangentially than it does radially (Figure 1.34).

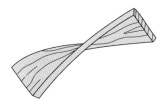

Figure 1.33 *Winding*

Shaking – also know as fissures (a separation of the wood fibre). These develop along the grain of a piece of wood, particularly at its ends, and are the result of drying too fast during seasoning. They are called 'splits' where the fissure extends through the board from side to side. They are called 'checks' if they are seen on the face or end grain but do not extend through to the other side (Figure 1.35).

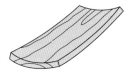

Figure 1.34 *Cupping*

Collapse – is also known as wash boarding and is caused by the cells collapsing through being kiln dried too rapidly (Figure 1.36).

Case hardening – This is the result of too rapid kiln drying. In this case the outside of the board is dry but moisture is trapped in the centre cells of the wood (Figure 1.37). This defect is not apparent until the board is re-sawn, when it will tend to twist. A simple test to confirm case hardening can be carried out on an end grain sample about 25 mm thick; the centre is cut out to form a two-pronged 'U' shape; the prongs closing up indicates case hardening.

Stick staining – is the result of using a different species of wood for the piling sticks to that being seasoned (Figure 1.38). It can be removed when the boards are processed.

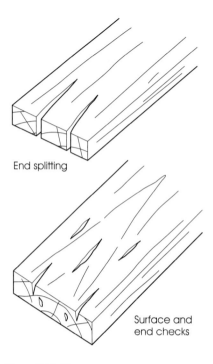

End splitting

Surface and
end checks

Figure 1.35 *Fissures*

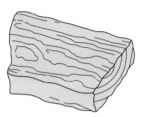

Figure 1.36 *Collapse*

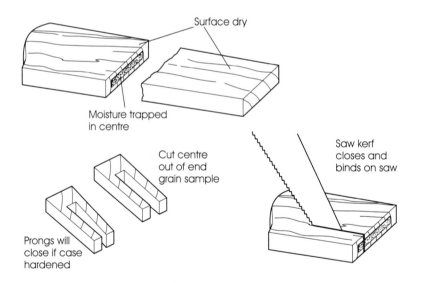

Surface dry

Moisture trapped
in centre

Cut centre
out of end
grain sample

Saw kerf
closes and
binds on saw

Prongs will
close if case
hardened

Figure 1.37 *Case hardening*

Stain left   Hardwood
by stick   stick

Softwood
board

Figure 1.38 *Stick staining*

did you know?

Sealing the ends of timber
with paint or nailed-on cover
strips can prevent splitting.

Natural defects

Shakes

Shakes are a separation in the wood fibre along the grain, which develop in the standing tree, on felling or prior to seasoning. The shake is formed as a result of stress relief, causing a longitudinal crack radiating from the heart and spreading along the length of the trunk (Figure 1.39).

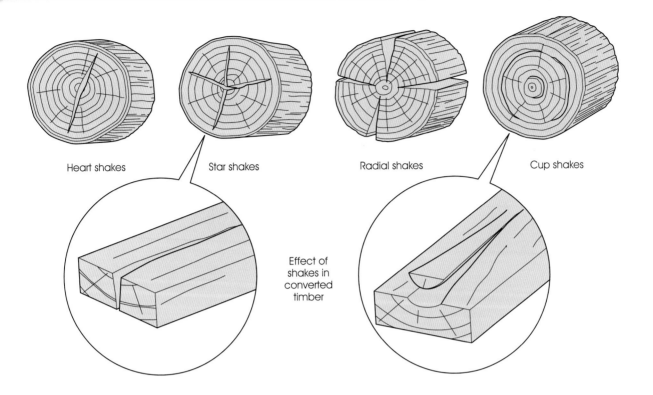

Heart shakes   Star shakes   Radial shakes   Cup shakes

Effect of shakes in converted timber

Figure 1.39 *Shakes*

Heart shakes – are splits along the heart of a tree and are probably due to over maturity.

Star shakes – are a number of heart shakes, which form an approximate star.

Radial shakes – are splits along the outside of the log, which are caused by the rapid drying of the outside of the log before it is converted.

Cup shakes – this is a separation between the annual rings and is normally the result of lack of nutriment. It is also said to be caused by the rising sap freezing during early spring cold spells.

Waney edge

This is where the bark is left on the edge of converted wood (Figure 1.40).

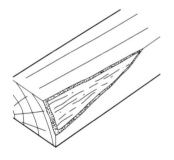

Figure 1.40 *Waney edge*

Knots

Knots are the end sections of branches where they grow out of the trunk (Figure 1.41). The presence of knots on the surface of a piece of wood often causes difficulties when finishing because of the distorted grain that knots cause.

Knots can be considered as being either live knots, i.e. knots that are firm in their socket and show no signs of decay (also known as sound or tight knots), or dead knots, i.e. knots that are separated from the surrounding wood by the bark of the branch, have become loose in their sockets or show signs of decay. Knots are also described by their size, location or number in a piece of wood (Figure 1.42).

did you know?

Dead or encased knots are the remains of branches, which have been overgrown by growth rings.

safety tip

Look out for dead, loose knots. They are a potential danger to the machinist as they may be picked up by the cutters and ejected rapidly towards the operator.

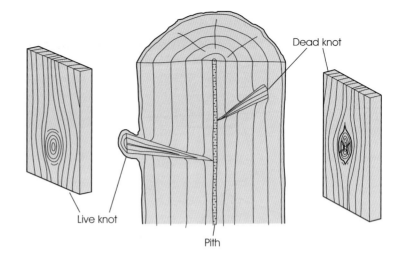

Figure 1.41 *Knots*

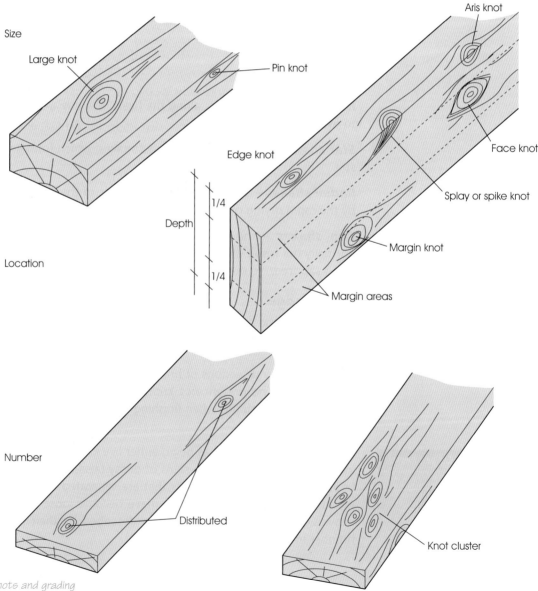

Figure 1.42 *Knots and grading*

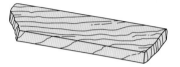

Figure 1.43 Upsets

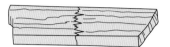

Figure 1.44 Sloping grain

Upsets

This is also known as thunder shake and is a fracture of the wood fibres across the grain (Figure 1.43). Upsets can be caused by the tree being struck by lightning during its growth, but they are mainly caused by the severe jarring the tree receives when being felled. This is a serious defect, which is most common in mahogany and is not apparent until the wood has been planed.

Sloping grain

This is where the grain does not run parallel to the edge of the board and is often caused by bad conversion (Figure 1.44). When the sloping grain is pronounced, the defect is called short graining. This seriously affects the strength of the wood and it should not be used for structural work.

Reaction wood

This is produced in timber that has had to grow in a leaning position either on a slope or against strong prevailing winds (Figure 1.45). The tree will attempt to produce extra growth on the trunk to counteract the force of gravity or wind force. Any extra growth will result in an eccentric appearance around the pith on the trunk cross-section. Softwoods put the extra growth on the compressed (lower or squashed) side of the trunk and is therefore termed compression wood. Hardwoods put the extra growth on the tension (upper or stretched) side and is termed tension wood. Reaction wood is prone to serious distortion during seasoning, its grain is often woolly and applied finishes may appear patchy.

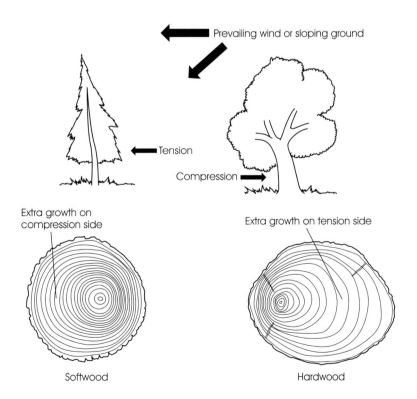

Figure 1.45 Reaction wood

did you know?

Sap stain is only normally considered as a defect in timber that is to have a polished or clear varnish finish.

Sap stain

Also known as blue sap stain or bluing, this often occurs in felled logs while still in the forest. It can also occur in damp wood that has been stacked close together without piling sticks or with insufficient air circulation. The stain is the result of a harmless fungus feeding on the contents of the sapwood cells. No structural damage is caused by this fungus.

Preservation

All timbers, especially their sapwood, contain food on which fungi and insects live. The idea behind timber preservation is to poison the food supply by applying a toxic liquid to the timber.

There are three main types of timber preservative available.

Tar oils – are derived from coal and are dark brown or black in colour. They are fairly permanent, cheap, effective and easy to apply. However, they should not be used internally, as they are inflammable and possess a strong lingering odour. They should never be used near food, as odour will contaminate it. Once treated, the timber will not accept any further finish, for example it cannot be painted. Traditionally, it has been used for the treatment of external timber such as fences, sheds and telegraph poles. However due to its carcinogenic risk (it can cause cancer) tar oil creosote is not available for public use and strict controls are in place for both its industrial use and the timbers treated with it.

Water-soluble preservatives – are toxic chemicals, mixed with water, and are suitable for use in internal and external situations. They are odourless and non-inflammable and the wood can be painted afterwards. As the toxic chemicals are water-soluble some of the types available are prone to leaching out when in wet or damp conditions.

Organic solvent preservatives – consist of toxic chemicals, mixed with a spirit that evaporates after the preservative has been applied. This is an advantage because the moisture content of the timber is not increased. The use and characteristics of these types of preservative are similar to those of water-soluble preservatives, but with certain exceptions.

safety tip

Preservatives can be toxic to you as well. Always follow the safety procedures.

Some of the solvents are flammable, so care must be taken when applying or storing them. Some types also have a strong odour. In general organic solvent preservatives are the most expensive type to use but are normally considered to be more effective.

Methods of application

did you know?

Pressure treatment is the most effective method of preservation

To a large extent it is the method of application rather than the preservative that governs the degree of protection obtained. This is because each method of application gives a different depth of preservative penetration. The greater the depth of penetration achieved, the higher the degree of protection. Preservatives can be applied using a number of methods but all of these can be classed in two groups:

Non-pressure treatment

This is where the preservative is applied by brushing, spraying or dipping. Brushing and spraying can be used for all types of preservative but the effect is very limited as they only give a surface coating. Dipping the timber into a tank of preservative gives a greater depth of penetration. This can be further increased by a process known as steeping. This involves heating the preservative with the timber in the tank. As it cools, the preservative gets sucked into the timber.

Pressure treatment

This is the most effective form of timber preservation as deep penetration is achieved. This process is carried out by suppliers with specialist compressed air or vacuum equipment. (Observe the difference in Figure 1.46.)

Little penetration surface coating only

Good sapwood but limited heartwood penetration

Complete sapwood and good heartwood penetration

Figure 1.46 *Depth of preservation penetration depends on method of application*

Brushing and spraying   Dipping and steeping   Pressure treatment

Preservative safety

◆ Always follow the manufacturers instructions with regard to use and storage.
◆ Avoid contact with skin, eyes or clothing.
◆ Avoid breathing in the fumes, particularly when spraying.
◆ Keep away from food to avoid contamination.
◆ Always wear personal protective equipment (PPE): barrier cream or disposable protective gloves; a respirator when spraying.
◆ Do not smoke or use near a source of ignition.
◆ Ensure adequate ventilation when used internally.
◆ Thoroughly wash your hands before eating and after work, using soap and water or an appropriate hand cleanser.
◆ In the case of accidental inhalation, swallowing or contact with the eyes, seek medical advice immediately.

Common woods and their main uses

The table below provides a checklist of wood types, names, sources and main uses with special features.

Name	Source	Colour	Main uses	Comments
European Redwood, also known as Scots pine, yellow deal or red pine (softwood)	Europe	Pale red to brown heartwood. Light yellow to brown sapwood	All general carpentry and joinery and structural work (roofs, floors, doors and windows)	Should be preservative treated when used externally. Also suitable for furniture and cabinet construction
Douglas fir, also known as British Columbian pine and Oregon pine (softwood)	Canada and USA	Pale brown	All general carpentry and joinery work (normally of a higher class than European Redwood), and plywood	Stains when in damp contact with iron. Also used for cabinet and furniture construction.
Western red cedar (softwood)	Canada and USA	Pink to red-brown	Interior and exterior joinery, cladding, panelling and roof coverings (shingles)	Stains when in damp contact with iron. Will weather to a silver grey colour
Whitewood, also known as Norway spruce and white deal (softwood)	Europe	White to light yellow	All general carpentry and structural work	Also suitable for external joinery if preservative treated
Parana pine (softwood)	South America	Light to dark brown with red streaks	Interior joinery and fittings	Straight grained, but prone to distortion.
Ash (hardwood)	Europe and USA	White, grey to light brown	Interior joinery and furniture	Has good bending properties. Also used for plywood and decorative veneers
Oak (hardwood)	Europe, America and Japan	Light brown with silver streaks	Panelling, furniture, external joinery, church fittings, floor and roof timbers	Hardwearing but stains when in damp contact with iron
Teak (hardwood)	South-east Asia	Mid-brown, often with greenish tints	High-class joinery, ship joinery, laboratory fittings	Extremely resistant to chemicals
Beech (hardwood)	Europe and Japan	Light brown with golden flecks	Furniture, plywood and floor coverings	Often steamed to produce a light pink colour
Mahogany (hardwood)	West Africa, Central and South America	Pink to red-brown	Panelling, cabinet work, flooring and joinery	Easy to work and finish

Purchasing wood

The terms 'wood' or 'timber' are taken to mean sawn or planed pieces in their natural or processed state. Different sections are available depending on their intended end use (Figure 1.47).

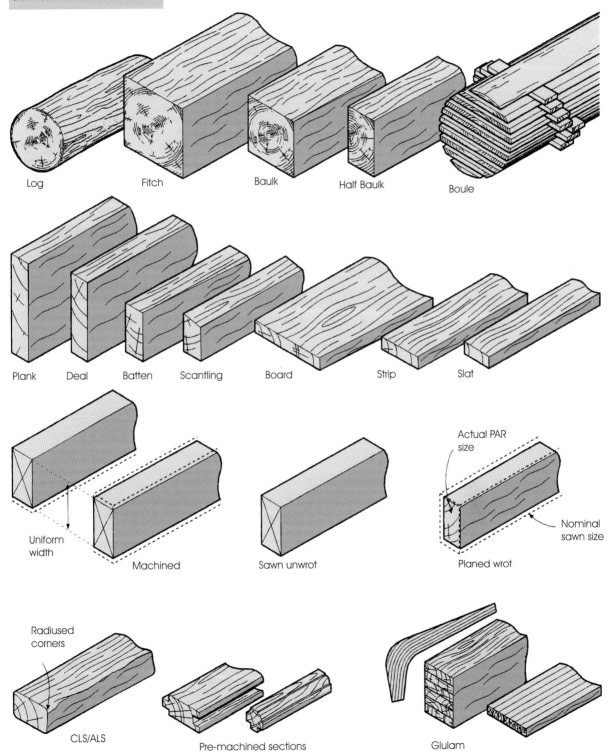

Log Fitch Baulk Half Baulk Boule

Plank Deal Batten Scantling Board Strip Slat

Uniform width Machined Sawn unwrot Planed wrot

Actual PAR size Nominal sawn size

Radiused corners

CLS/ALS Pre-machined sections Glulam

Figure 1.47 *Standard timber sections*

Materials Chapter 1

Log

Normally a debarked tree trunk.

Main uses:

- Hardwoods are often shipped in this form for conversion in UK saw mills.
- Poles or posts for agricultural, garden, dock and harbour work.

Flitch – large square log.

Baulk – squared log over 100 mm × 125 mm.

Half baulk – any baulk sawn through its section in half.

Main uses:

- Re-conversion into smaller sections.
- Dock and harbour work, shoring, hoarding, beams and posts in heavy structural work.

Boule

Through and through sawn log re-stacked into its original log form. Each piece having waney edges.

Main uses:

- Mainly hardwood for re-conversion into smaller sections.
- Rustic cladding and garden furniture.

Sawn or planed sections

Sawn wood may be termed as unwrot (unworked) and planed as wrot (worked). These may be further defined as:

Plank – 50 mm to 100 mm in thickness and 250 mm or more in width.

Deal – 50 mm to 100 mm in thickness and 225 mm to 300 mm in width. Also used to refer to a species of timber, e.g. red deal for redwood.

Batten – 50 mm to 100 mm thick and 100 mm to 200 mm in width. Also a term loosely used in the trade for timber up to 25 mm × 50 mm in section and, in certain parts of the country, to refer to a scaffold board.

Scantling – 50 mm to 100 mm thick and 50 mm to 125 mm in width.

Board – under 50 mm thick and over 100 mm in width.

Strip – under 50 mm thick and under 100 mm in width.

Slat or slatting – 25 mm and under in thickness and 100 mm and under in width.

Die squared – a baulk dimension stock wood that has been converted to standard sizes.

Machined – piece of wood mainly used in carpentry, which has been planed or sawn to a regular width (depth) from one or both edges.

Nominal – the sawn or 'ex' size of a piece of planed wood. The actual PAR size will be smaller.

CLS/ALS – Canadian lumber stock or American lumber stock processed wood with rounded corners for structural work, 38 mm thickness, and widths from 63 mm to 285 mm.

Main uses:

◆ Sawn and machined sections for carcassing, first-fixing and other general construction work. Also for joinery, furniture and cabinet work for re-conversion and further processing.
◆ Planed sections: finished sections in carpentry first and second fixing also in joinery, furniture and cabinetwork for re-conversion and/or further processing.

Pre-machined sections

Planed sections that have been further processed for a particular end use.

Main uses:

◆ Carpentry flooring, cladding, skirting, architrave and window boards etc.
◆ Joinery sections for window, door and other components.
◆ Picture frames and other general moulding for furniture and cabinet work.

Glue-laminated timber – also known as glulam

Main uses:

◆ Structural work especially where large sections, long spans or shapes are required.
◆ Joinery, furniture and cabinet work, especially for unit construction, shelving, counter tops, table tops and other work tops.

Common abbreviations

There are many abbreviations used when describing timber. These are used particularly in price lists and trade catalogues. Some terms are illustrated in Figure 1.48.

AD	air dried
ak	boards re-sawn after kiln drying
avge	average
bd	board
bdl	bundle
com	common
com. & sels	common and selects
clr & btr	clear and better
DAR	dressed all round
d.b.b.	deals, battens, boards (sizes of timber)
ex	nominal size

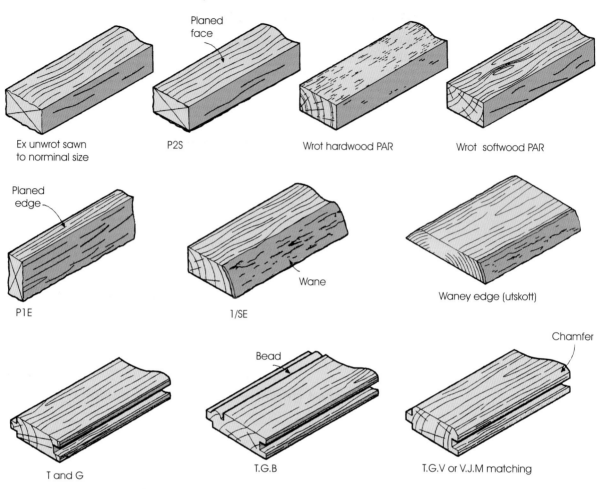

Figure 1.48 *Frequently used terms and abbreviations*

FAS	firsts and seconds
FSP	fibre saturation point
hdwd	hardwood
HG or BG	home grown or British grown
KD	kiln dried
lgth	length
MC	moisture content
merch.	merchantable
P1E or S1E	planed or surfaced one edge
P2E or S2E	planed or surfaced two edges
P1S or S1S	planed or surfaced one side
P2S or S2S	planed or surfaced two sides
P1S1E or S1S1E	planed or surfaced one side and one edge
P2S1E or S2S1E	planed or surfaced two sides and one edge
P1S2E or S1S2E	planed or surfaced one side and two edges

P4S or S4S	planed or surfaced four sides
PAR	planed all round
PE	plain edged
PHND	pin-holes no defect (a grading term indicating that pin-holes are not considered to be a defect)
PSE	planed and square edged
PSJ	planed and square jointed
PTG	planed, tongued and grooved
qtd	quartered
sap	sapwood
S/E	square edged boards
1/SE	a board with one square and one waney edge
sftwd	softwood
sels	selects
SND	sapwood no defect (a grading term indicating that sapwood is not considered to be a defect)
T&G	tongued and grooved
T/T	through and through sawn
TGB	tongued, grooved and beaded
TGV	tongued, grooved and V-jointed
U/E	unedged
U/S	unsorted
VJM	V-jointed matching
W/E	waney edged
WHND	worm-holes no defect (a grading term indicating that worm-holes are not considered to be a defect)
wt	weight

Available sizes

Softwoods

Sawn softwoods are available in a wide range of preferred sectional sizes (the most usual sizes available in Europe) as illustrated in Figure 1.49, which shows the basic sizes (sawn size) at a moisture content of 20%.

Machined softwoods – are sawn or planed on one or both edges to give a consistent depth for structural use (such as floor joists) and have a target size (the desired size after a production process) of 3 mm less than the sawn size for sections up to 150 mm deep, and 5 mm less for those over 150 mm in depth (Figure 1.50).

Materials

Chapter 1

Figure 1.49 *Typical range of sawn softwood sizes*

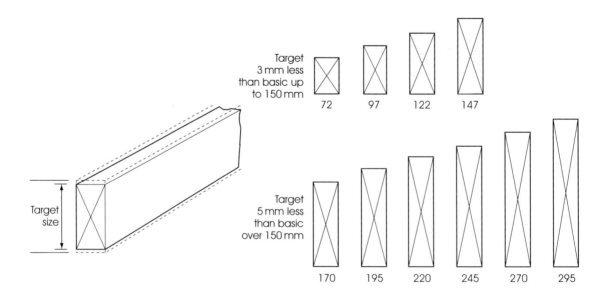

Figure 1.50 *Target sizes for structural softwoods*

Surfaced timber – known as CLS or ALS (Canadian or American Lumber sizes/stock). It is a planed timber with rounded arrises of up to 3 mm radius for ease of handling, which is used for structural purposes including joists, timber frame panels and stud partitioning. Available in a thickness of 38 mm and widths of 63 mm, 89 mm, 114 mm, 140 mm, 184 mm, 235 mm and 285 mm.

Planed or prepared timber – is sawn timber that has been machined for its full length and width on at least one face to provide a smooth surface. Prepared timber may also have been cut to length. Normally referred to by its ex (out of) nominal or sawn size. The following reductions apply to the sawn section:

◆ 3mm less than the basic sawn section up to 150 mm
◆ 5mm less than the basic sawn section over 150 mm.

Softwood lengths are in 300 mm increments starting at 1.8 m up to 7.2 m, although lengths above 5.7 m are limited and lengths of 6 m and over may not be available without finger jointing (Figure 1.51). No minus tolerance in length is permitted; however, plus tolerances can be agreed at time of order.

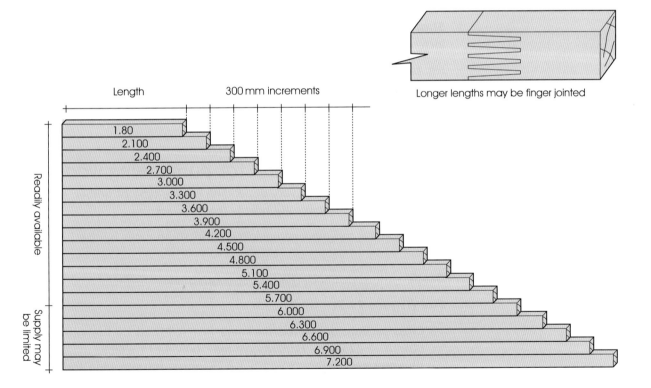

Longer lengths may be finger jointed

Length   300 mm increments

1.80
2.100
2.400
2.700
3.000
3.300
3.600
3.900
4.200
4.500
4.800
5.100
5.400
5.700
6.000
6.300
6.600
6.900
7.200

Readily available

Supply may be limited

Figure 1.51 *Available softwood lengths*

Materials

Chapter 1

Hardwoods

Hardwoods are quoted in metric or imperial thickness depending on source.

$\frac{1}{2}''$	or	13 mm
$\frac{3}{4}''$	or	19 mm
1"	or	25 mm
$1\frac{1}{4}''$	or	32 mm
$1\frac{1}{2}''$	or	38 mm
2"	or	50 mm
$2\frac{1}{2}''$	or	63 mm
3"	or	75 mm
4"	or	100 mm

Thicker sections may be available, rising in 10 or 25 mm stages. Widths are in accordance with the grade, normally 3" or 75 mm and wider, rising in increments of 10 or 25 mm.

Normal lengths – are 1.8 m and above rising in 100 mm increments.

Short lengths or shorts – are 1.7 m and less falling by 100 mm increments.

Figures 1.52 and 1.53 illustrate standard sizes of hardwoods.

The width of waney-edge pieces are measured at three points along the narrow face including half the wane. The stated width is the average of the three measurements (Figure 1.54).

Information required for purchasing wood

Information required will vary depending on the type of timber, end use and supplier, but will typically include the following:

Species – the common name followed in some cases by the botanical name.

Grade – the appearance grade or strength grade.

Size (Figure 1.55) –

◆ Cross section: quote the thickness first followed by the width, e.g. 50 mm × 225 mm. Hardwoods may only be available in a stated thickness and minimum width, e.g. 50 mm × 100 mm wide and up.
◆ Volume: larger quantities of both hardwoods and softwoods are purchased by the cubic metre (m³); hardwoods may also be measured by the cubic foot.
◆ Area: floor-boarding and cladding is purchased by the square metre (m²).
◆ Length: pre-machined mouldings, skirtings and architraves, etc. are purchased by the running metre (m/run).

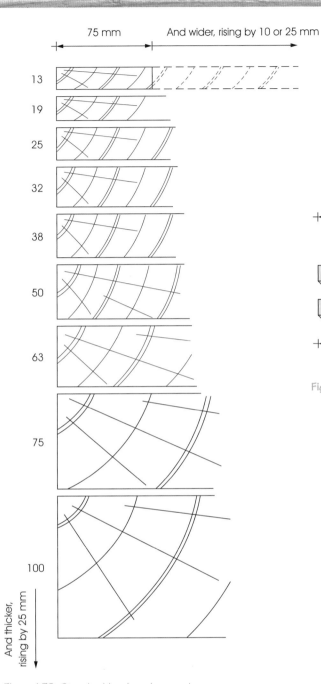

75 mm · And wider, rising by 10 or 25 mm

13
19
25
32
38
50
63
75
100

And thicker, rising by 25 mm

Figure 1.52 *Standard hardwood sawn sizes*

Shorts up to 1.700

And less
falling by 100 mm

Normal 1.800 · And above

Rising by
100 mm

Figure 1.53 *Standard hardwood lengths*

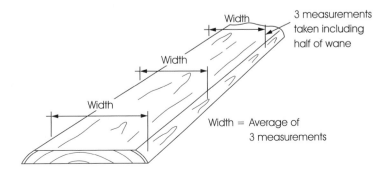

Width

Width

Width

3 measurements
taken including
half of wane

Width = Average of
3 measurements

Figure 1.54 *Measuring waney-edge boards*

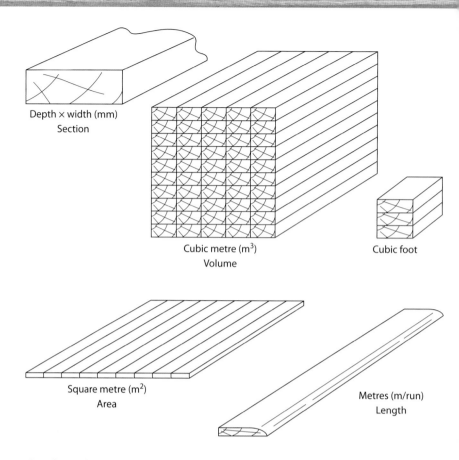

Depth × width (mm)
Section

Cubic metre (m³)
Volume

Cubic foot

Square metre (m²)
Area

Metres (m/run)
Length

Figure 1.55 *Methods of purchasing wood*

Finish – the type of finish or processing required (sawn, regularised, planed or moulded).

Moisture content – the moisture content or the method of seasoning if required.

Preservative treatment – the type of treatment if required.

> *example*
>
> **European redwood U/S, 50 × 225, 2m³ PAR KD**

Grading of wood

Wood is available graded to give an indication of its appearance or strength. Softwoods and hardwoods are graded in different ways and the rules governing grading systems vary from country to country.

Softwood appearance grading

Softwood from European sources are generally graded into six numbered groups, I, II, III, IV, V and VI, according to the amount of defects they contain. First being the best quality through to sixth being the poorest quality (Figure 1.56).

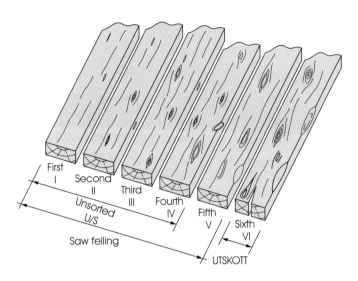

Figure 1.56 *Softwood appearance grades*

◆ Grades 'I, II, III and IV' are rarely available separately. They are normally marketed together as U/S or unsorted, providing a mixed batch of grades I to IV. This is the normal grade specified for joinery work.
◆ The 'V' grade is available separately and is used for general non-structural construction and carcassing work.
◆ The 'VI' grade is a low quality wood used mainly for packaging; it is available separately and is often known as 'UTSKOTT', a general term meaning waney-edged boards.
◆ Other parcels of timber may be described as saw felling quality. This is a mixed batch of grades I to V sold without further sorting, but the majority will consist of IV and V grades with very little of I to III grades. Some suppliers market this as 'V and better' or 'fifths and better'.

Countries of the former Soviet Union use a similar system, but only have five numbered grades: I, II and III are marked as unsorted; IV and V are equivalent or slightly better than the European V and VI.

Canadian and North American grading rules use the term 'clear' for their best grade, which is virtually free of defects. These are equal to or better than European I grade. 'Selected merchantable' and 'merchantable' are the middle grades with 'commons' as the lowest grade and equivalent to European grade VI.

Wood graded in other parts of the world may use other grading rules and terms; check with the supplier with regards suitability for your intended end use.

Shipping marks – Exporting saw mills mark the cut ends of every piece of wood with their own private grade of shipping mark. They may be stamped, stencilled or branded using a combination of letters, symbols and colours. Knowledge of these can identify its origin and grade. Details can be found in a directory of shipping marks. Examples are shown in Figure 1.57.

did you know?

You can consult a 'Directory of Shipping Marks' in many libraries to find details of the timber you are working with.

Hardwood appearance grading

Hardwoods come from a far wider range of sources than softwoods, resulting in more varied grading systems. Often a grading will not be

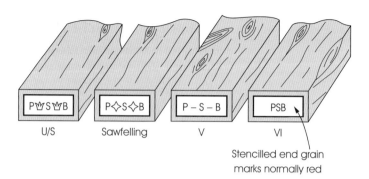

P✿S✿B   P◇S◇B   P – S – B   PSB

U/S   Sawfelling   V   VI

Stencilled end grain
marks normally red

*Figure 1.57 Typical shipping
marks and grades*

applied and instead the wood is sold for a specific end use, by mutual agreement between the buyer and seller.

Most hardwood grades are based on the usable area of wood that is obtainable from a board. The greater the area the higher the grade. This is known as the cutting system (Figure 1.58).

- ◆ The best grades 'first and seconds' (FAS) provide long, wide, clear cuttings of at least 83.4% of the board from the worst face. The other face will be at least the same or better. They are ideally suited for furniture, high quality joinery and mouldings.
- ◆ 'First and seconds one face' (FAS 1F) and 'selects' grade the board from both faces. The best face must meet the requirements for FAS, the other meet No. 1 common requirements. The difference between FAS 1F and selects is the cutting's minimum length and width requirements (typically 2440 mm by 150 mm for FAS 1F and 1830 mm long by 100 mm wide for selects).
- ◆ No. 1 commons provide clear cuttings of medium length and width from at least 66.7% of the boards (typically at least 900 mm long by 75 mm wide or 600 mm long by 100 mm wide).
- ◆ No. 2 commons provide clear cuttings of short length and narrow width from at least 50% of the board's area (typically at least 600 mm long by 75 mm wide).

Both common grades are an economical option for smaller joinery components, furniture, framework and a range of smaller wooden items.

Parcels of timber may contain a mixture of grades. No. 1 commons may be mixed with FAS 1F boards and sold as 'No 1 commons and better'. The expected percentage of better grades will be identified by the supplier.

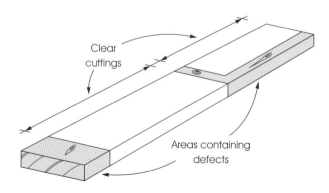

Clear
cuttings

Areas containing
defects

*Figure 1.58 Hardwood appearance
grading based on
cutting system*

Strength grading

Both softwoods and hardwoods are available in strength grades (formerly known as stress grades). There are two methods used for strength grading:

Visual strength grading – Each piece is inspected and either rejected or given a grade, taking into account the number, size and positioning of defects.

Machine strength grading – Each length is passed through a machine, which deflects the piece to test its stiffness. Based on the stiffness, a grade is automatically marked on the face of each piece of wood.

Softwood grades are class 'C': C14, C16, C18, C22, C27, C30, C35, and C40. Hardwood grades are class 'D': D30, D35, D40, D50, D60, and D70. The numbers indicate a characteristic value of strength and stiffness. The higher the number, the stronger the piece. Figure 1.59 illustrates typical strength grading marks.

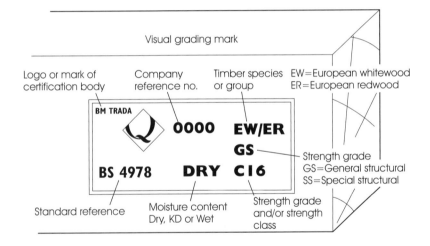

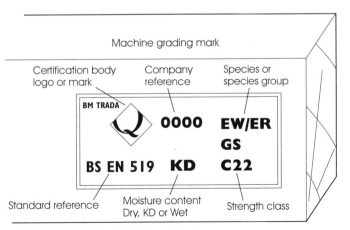

Figure 1.59 *Typical strength grading marks*

measuring up

1. Which of the following timber is most commonly used for carpentry?
- **(a)** oak
- **(b)** Parana pine
- **(c)** spruce
- **(d)** European redwood

2. Describe THREE main differences between softwoods and hardwoods.

3. Softwood is available in stock lengths from 1.8 m. State the measurement that stock lengths increase by.

4. What is the actual cell-producing part of a tree is known as?
- **(a)** bark and bast
- **(b)** sapwood
- **(c)** cambium layer
- **(d** medullary rays

5. Define the term 'conversion' as used in the timber trade.

6. Make a sketch to show a quarter sawn board.

7. Define the term 'nominal', when referring to the size of a piece of wood.

8. State what is meant by the abbreviation PAR.

9. Name TWO types of timber preservative and state TWO methods of application.

10. Which of the following methods of applying timber preservative attains the deepest penetration?
- **(a)** spraying
- **(b)** dipping
- **(c)** pressure impregnation
- **(d)** brushing

Wood-based board materials

There is a wide range of wood-based board materials available to the woodworker. These generally fall into three main categories:

- ◆ laminated boards;
- ◆ particleboards;
- ◆ fibreboards.

Chapter 1 Materials

Laminated boards

Plywood

Plywood is a laminated board material made from thin sheets of wood, termed construction veneers or plies. Its panel size, stability and ease of use, make it an ideal material for both construction work (flooring, cladding and formwork) and interior joinery and carcass work (panelling, furniture and cabinet construction). (See Figure 1.60.)

Plywood is normally constructed with an odd number of layers, with their grain direction alternating across and along the sheet to counter movement in the wood. When glued together it forms a strong board that will retain its shape and not have a tendency to shrink, expand or distort. The number of plies varies according to the thickness of the finished board, three being the minimum. Whatever the number of plies, the construction is symmetrical about the centre ply or centreline of the board, so that both sides are equally balanced to resist warping. Where an even number of plies are used, the central two plies are bonded with their grain running parallel to each other, to act as a single layer and thereby maintain the balance of the board.

Plywood sizes

Plywood is manufactured in a range of sizes, varying in thickness from 3 mm to 30 mm, in increments of approximately 3 mm. The width of the board is typically 1.22 m but 1.52 m boards are also available.

The length of the most common board is 2.44 m, but boards up to 3.66 m are manufactured. The grain direction of the face plies normally run the longest length of the board. However short grain boards are also available. Thus when specifying plywood it is standard practice for the grain direction to run parallel to the first stated dimension. Therefore a 2.44 m × 1.22 m board will have long grain, while a 1.22 m × 2.44 m board will have a short grain (Figure 1.61).

did you know?

Plywoods and other decorative veneered boards may be described as having either quarter or crown cut faces. Quarter cut veneers are radially sliced resulting in a straight grained board; whereas crown cut veneers are tangentially sliced to give a characteristic flame or arched top figured board.

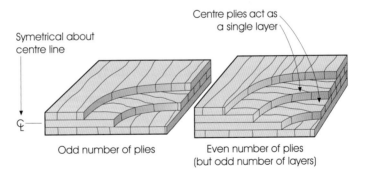

Symetrical about
centre line

Centre plies act as
a single layer

Odd number of plies

Even number of plies
(but odd number of layers)

Figure 1.60 *Plywood symmetrical construction*

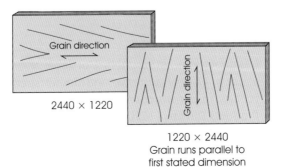

Grain direction

2440 × 1220

Grain direction

1220 × 2440
Grain runs parallel to
first stated dimension

Figure 1.61 *Plywood sizes and grain direction*

Materials   Chapter 1

Appearance grading

Plywood is graded according to the appearance of its outer faces, each face being assessed separately. In common with timber, the grading rules for plywood vary widely from country to country. Most plywood manufacturers base their grading rules on a coding system to indicate board grades.

Northern Europe/Russia – These grades are listed with their permitted defects:

A   practically defect free

B   a few small knots and minor defects

BB   several knots and well-made plugs

C or WG   all defects allowed: only guaranteed to be well glued

It is rarely necessary for the face and back veneer to be of the same grade. Most manufacturers offer a wide combination of face/back veneer grades: A/A A/B A/C B/B B/C, etc. The back veneer may also be described as a balancer where it will not be seen and need not be from the same species of wood. These boards may be graded as A/bal or B/bal. Some suppliers of veneered boards term A/A or A/B boards as double-sided DS and A/bal or B/bal boards as single-sided SS.

Canada/America – Plywood from these sources are available in the following grades, which are approximately equivalent to the Northern Europe and Russian grades shown.

- G2s (good two sides) equivalent to A/A
- G/S (good one side/solid reverse) equivalent to A/B
- G1s (good one side) equivalent to A/C
- Solid 2s (solid two sides) equivalent to B/B
- Solid 1s (solid one side) equivalent to B/C
- Sheathing equivalent to C/C

Bonding

The performance of plywood is dependent not only on the quality of the plies used, but also on the type of adhesive used in manufacture.

- INT (interior grade) will not withstand humidity or dampness.
- MR (moisture resistant) and BR (boil resistant) are suitable for use under normal conditions, but will not withstand continuous exposure to extreme conditions.
- WBP (weather and boil proof) are suitable for use under any conditions.

See Adhesives later in this chapter for further details.

Labelling

Manufacturers either stencil or label their products to indicate the face veneers, type of adhesive and timber species. Plywoods and other panels for use in construction have to comply with the European Construction Products Directive (CPD), and may have a 'CE' label on them. Typical labels are illustrated in Figure 1.62.

Typical label information

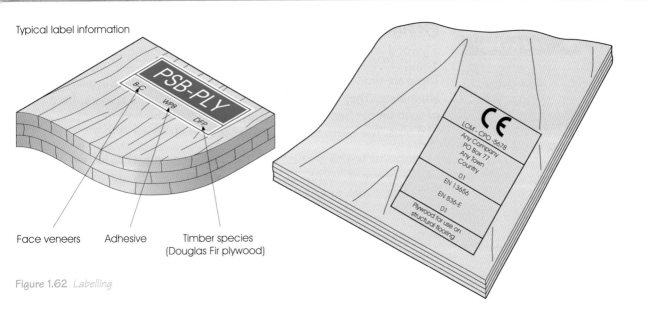

Face veneers   Adhesive   Timber species
(Douglas Fir plywood)

Figure 1.62 *Labelling*

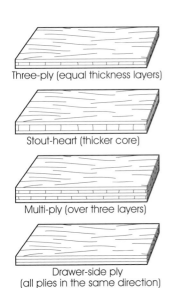

Three-ply (equal thickness layers)

Stout-heart (thicker core)

Multi-ply (over three layers)

Drawer-side ply
(all plies in the same direction)

four- and six-ply
(central plies lay in same direction)

Figure 1.63 *Types of plywood*

Types of plywood

Plywood boards are manufactured in many parts of the world, and thus the species of wood used varies dependent on their origin. The core and face plies may be from the same wood throughout or of different species.

◆ Softwood boards are normally from Douglas fir.
◆ Hardwood boards using light coloured temperate woods are normally from either birch or beech.
◆ Tropical hardwood boards using a variety of darker red colour woods, including Gaboon, Lauan and Meranti.

Types of plywood (see Figure 1.63) include the following:

Three-ply – consists of three equal thickness layers; ideal for drawer bottoms and cabinet backs.

Stout-heart – also consists of three layers, but the middle layer is thicker.

Multi-ply – is the name given to any plywood which has more than three layers. Softwood boards are used for formwork and structural work, while hardwood boards are used for furniture and cabinet construction.

Drawer-side ply – is the exception to the alternating cross grained construction as all plies run in the same direction. It is normally made from birch in a thickness of 12 mm and is used in place of solid timber for drawer construction.

Four-ply and six-ply – both have their central plies bonded together with their grain in the same direction; mainly manufactured from softwood for structural use.

Decorative-ply – made from multi-ply, which has been faced with selective crown or quarter cut matching veneers; used for panelling, furniture and cabinetwork. To maintain the stability of the board, a balancer, of lesser quality veneer, must be applied to the reverse face of the board. These may be graded as A/B, A/C or A/balancer.

Other laminated boards

Laminboard and blockboard

These are similar to ply, being a layered construction with the grain direction alternating. They differ from plywood in that the core is made from strips of solid timber (see Figure 1.64). These are faced on both sides with one or two plies. The width of the strips varies with each type of board.

Laminboard has strips that are up to 8mm. In blockboard the strips are up to 25mm in width. Both forms of laminated board are available in the same panel sizes as plywood. Thicknesses vary between 12mm and 38mm.

Both are used as a core material for veneering, panelling, partitioning, door-blanks, furniture and cabinet work. Laminboard is superior to blockboard, as there is less likelihood of the core strips distorting and showing through on the surface of the board – a defect termed 'telegraphing'.

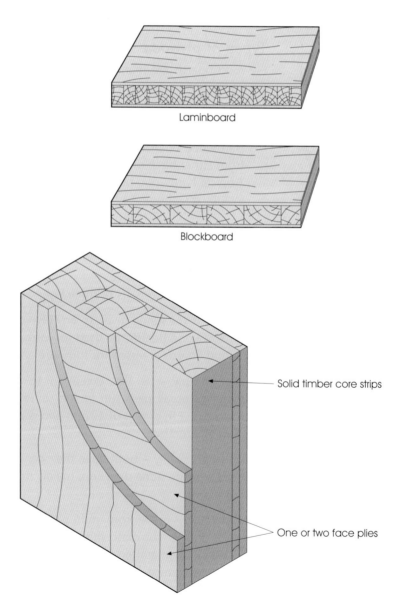

Laminboard

Blockboard

Solid timber core strips

One or two face plies

Figure 1.64 *Solid core laminated boards*

Particleboards

Wood particleboards are manufactured using small wood chips or flakes impregnated and bonded together under pressure. Softwood forest thinnings and waste from sawmills and other wood manufacturing processes are mainly used.

Particleboard is manufactured in a range of sizes, varying in thickness from 2.5 mm to 38 mm with extruded tubular core boards increasing to 50 mm or more. The standard imperial 'eight by four' board is still readily available as 2440 mm × 1220 mm. The fully metric equivalents are 2.4 m × 1.2 m boards based on 600 mm increments. Lengths range from 1.83 m up to 3.66 m. Boards 600 mm wide are made mainly for flooring use.

Chipboard

Types of chipboard are shown in Figure 1.65. If exposed to excess moisture, the board swells considerably in thickness, loses strength and does not subsequently recover on drying. Stronger, moisture-resistant grades, tinted green in colour, are made for flooring and for use in damp conditions. Red tinted boards are also available for where fire resistance is required.

Single-layer chipboard – is made from a mass of similar size wood particles, pressed to form a flat, relatively coarse surface board. This is suitable as a core board for veneering, or plastic laminating, but not painting.

Three-layer chipboard – consists of a low density core of large particles, sandwiched between two outer layers of finer, higher density particles. The outer layers contain a higher proportion of resin. This produces a very smooth, even, surface that is suitable for painting. Extra care is required when sawing layered boards, as, compared to single layered, they tend to split or de-layer at the cut edge.

Graded density chipboard – has a core of coarse particles and faces of very fine particles. But, unlike the layered boards, there is a gradual transition from the coarse core to the fine surface. These are ideal for use in furniture and cabinet construction.

Extruded boards – are formed by forcing a blend of wood particles and resin through a die, resulting in a continuous length of board to the required width and thickness. Most particles are at right angles to the board face, thus reducing strength. The holes in tubular core boards are formed by heating coils used in the curing process. This enables much thicker boards to be made, as well as reducing the overall weight. Their main uses are for partitioning, door blanks and core stock for veneers and melamine foils.

Decorative-faced chipboard – normally a single layer, graded density or extruded board faced with a wood veneer, plastic laminate or a thin melamine foil (MFC, melamine-faced chipboard). The veneered boards are supplied sanded, ready for polishing. No further surface finishing is required for the plastic laminated or melamine foil-faced boards. These are used extensively for furniture, shelving and cabinet construction. The plastic laminate-faced boards often have a thicker core and a post-formed rounded edge, for use as kitchen worktops.

did you know?

In general the denser the chipboard, the stronger it is.

Single-layer

Three-layer

Graded density

Extruded

Extruded tubular core

Figure 1.65 *Types of chipboard*

Materials

Chapter 1

Oriented strand board (OSB)

This is manufactured from large softwood strands bonded with resin and wax in three layers (Figure 1.66). The strands in each layer are oriented in the same direction, with face layers being at right angles to the core layer, much like three-core plywood. Main uses are as floor and roof coverings, wall sheathing, site hoardings, formwork and packing cases.

Flake-board or wafer-board

Consists of wood flakes or shavings often hardwood. The flakes are laid horizontally and overlap each other, with their grain direction orientated randomly. These have a similar range of uses to OSB.

Fibreboards

These are manufactured from pulped wood and other vegetable fibre, pressed into boards of the required thickness. Boards of varying density are produced depending on the pressure applied and the process used (Figure 1.67).

Hardboard is available in thickness from 1.5 mm to 12 mm, medium and soft boards from 6 mm to 12 mm thick and MDF from 1.5 mm to 60 mm thick. Board sizes in general vary from 2440 mm to 3660 mm in length and 1220 mm to 1830 mm in width. Smaller and larger sizes may be available for certain boards to special order.

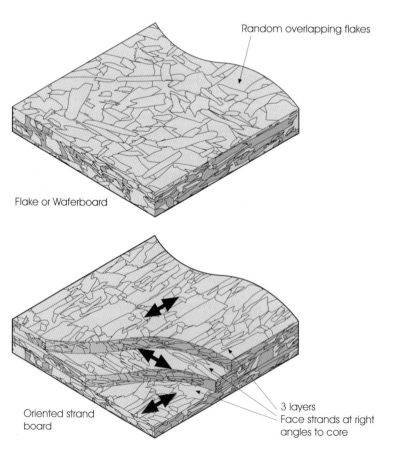

Random overlapping flakes

Flake or Waferboard

Oriented strand board

3 layers
Face strands at right angles to core

Figure 1.66 *Other particle boards*

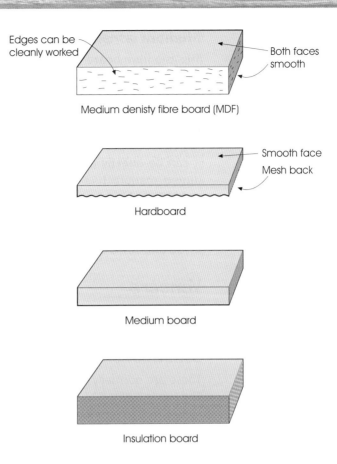

Figure 1.67 *Types of fibreboard*

Hardboard

Hardboard is high-density fibreboard made from wet wood fibres pressed together at high pressure and temperature. This wet process uses the natural resins in the wood fibres to bond them together.

Standard hardboard – has a smooth upper face, the opposite face is rough, caused by the meshed lower press plate that allows drainage of the moisture.

Duo-faced hardboard – has two smooth faces and is produced by further pressing standard boards between two flat plates.

Tempered hardboard – is a standard board that has been impregnated with oil and resin to produce a stronger board that is more resistant to moisture and abrasion.

Decorative hardboard – is produced in boards which are moulded, perforated, wood grained, painted and melamine foil faced.

Medium board

This is made in the same way as hardboard, using a wet process, but lower pressure, resulting in a mid-density board.

Soft board or insulation board

Soft board is a low-density board made using the wet process, but only lightly compressed and dried in an oven. Bitumen-impregnated soft boards are available for where increased moisture resistance is required.

did you know?

Decorative veneers are often applied to the faces of particle and fibre boards. These boards will then be graded in a similar way to plywood, e.g. A/A or A/B. In situations where only one face board will be seen in the finished job, the unseen face must still be overlaid with a balancing veneer to maintain the board's stability, otherwise it will distort.

Materials Chapter 1

Medium density fibreboard (MDF)

This is a fibre board with two smooth faces, but unlike other fibre boards it is made using a dry process with synthetic resin adhesive added to bond the fibres together. MDF can be worked with hand and machine tools like solid wood. It has a fine texture that allows faces and edges to be cleanly worked and finished. It provides an excellent core board for veneers, plastic laminates and melamine foil (MFMDF). In addition, it can be directly painted or polished. Standard boards can be modified during manufacture to increase their moisture or fire resistance. Green tinted boards are moisture resistant (MRMDF). Red tinted boards are fire resistant (FRMDF). MDF is widely used for panelling, furniture, cabinet construction and joinery sections such as mouldings, skirting, architraves, fascia and soffit boards.

Veneers

Veneers are very thin sheets or 'leaves' of real wood, which are cut from a log for decorative or constructional purposes (Figures 1.68 and 1.69).

Constructional veneers

These are normally rotary peeled. The whole log is put into a giant lathe. The log is rotated against a knife, which runs the whole length of the log to peel off a continuous sheet of veneer. A widely varying grain pattern is produced which is not normally considered decorative. Thus these veneers are mainly used for plywood manufacture.

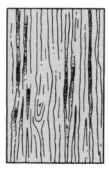 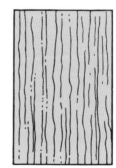 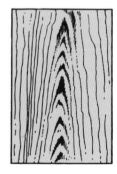

Figure 1.68 *Veneer cutting and matching*

Rotary peeled   Radially sliced   Tangentially sliced

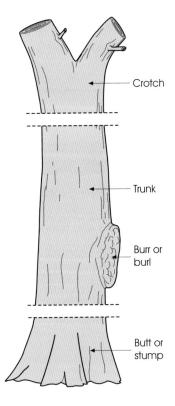

Figure 1.69 *Tree parts used for veneers*

Labels on figure: Crotch, Trunk, Burr or burl, Butt or stump

Decorative veneers

Sliced either radially or tangentially to provide either quarter cut (QC) or crown cut (CC) figuring. These 'leaves' are used for facing plywood and other decorative boards using book or slip matching. Most decorative veneers are sliced at 0.6 mm thick. Thicker veneers are normally available in lengths of 2440 mm and longer. Leaf widths of tangential sliced veneers range from about 225 mm to 600 mm depending on the species. However radial sliced leaves tend to be much narrower.

Once sliced and dried, the veneers are kept in multiples of four leaves and bundled into parcels of 16 to 32 leaves for matching purposes. Each bundle or parcel is taped together and re-assembled in consecutive order into its log form, ready for use.

More exotic figured veneers are available by slicing from the butt, burr or crotch of a tree. These exotic pieces are highly prized for furniture and small wooden articles. They are normally much shorter in length or may be of an irregular shape. Burrs for example are irregular and typically from 150 mm × 100 mm to 1000 mm × 450 mm in size.

Laminated plastic

This is a thin synthetic plastic sheet, made from about ten layers of craft paper bonded and impregnated with a resin. A plain colour, patterned or textured paper or real wood veneer is used for the upper decorative face, which is coated with a clear coat of melamine formaldehyde to provide a hardwearing, hygienic, matt, satin or gloss finish (Figure 1.70). Plain non-face sheets are also available for use as backing laminates.

Thickness ranges from 0.6 mm to 1.5 mm. Lengths range from 1830 mm to 3660 mm and widths from 600 mm to 1830 mm. The thinner sheets are used for post-forming and curved work and the thicker ones for flat working surfaces.

Purchasing board material

Board material is normally sold by the sheet, but may be priced by the square metre (m²), by 10 square metres (10 m²) or by the individual sheet. Typical information required when purchasing board material is:

- thickness of board;
- type of board;
- board finish;
- size of board – quote longest grain size first if applicable;
- number of boards.

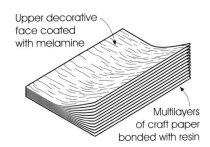

Figure 1.70 *Laminated plastic*

Labels on figure: Upper decorative face coated with melamine; Multilayers of craft paper bonded with resin

> **18 mm MDF Cherry (CC) veneered A/B 2400 × 1200 mm 25 No. off.**

Woodscrews

Woodscrews are used for joining wood to wood. The clamping force they provide makes a strong joint that can be easily dismantled. Screws are also used to attach items of ironmongery, such as hinges, locks and handles.

Most screws are made of steel. They may be treated to resist corrosion and case hardened for additional strength. Chrome plated or black japanned steel screws are used for decorative purposes. Brass screws are more decorative and stainless steel screws are more resistant to corrosion. Both brass and stainless steel screws can be used with acidic hardwoods such as oak, which are stained by steel screws.

Screw sizes

The size of a screw is specified by the length of the part that enters the wood and the diameter of the shank (Figure 1.71). Metric shank sizes are in millimetres and imperial sizes indicated by a gauge number (Figure 1.72).

Screw threads

Conventional woodscrews

These are threaded for about 60% of their length. The plain shank acts as a dowel, and the larger head holds the two pieces of wood or the item of ironmongery in place.

Twin-threaded woodscrews

These have a coarser twin thread that provides a stronger fixing in wood-based boards such as chipboard and MDF compared with conventional woodscrews. More of the screw length is threaded and the steeper pitch enables the screw to be driven in quicker. In addition the shank is narrower to reduce the risk of splitting (Figure 1.73).

did you know?

There are no direct equivalents between the shank size of imperial and metric screws, but the following are close:

No. 4 = 3.0 mm

No. 6 = 3.5 mm

No. 8 = 4.0 mm

No. 10 = 5.0 mm

No. 12 = 6.0 mm

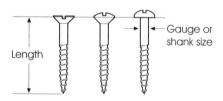

Length

Gauge or shank size

Figure 1.71 *Screw size data*

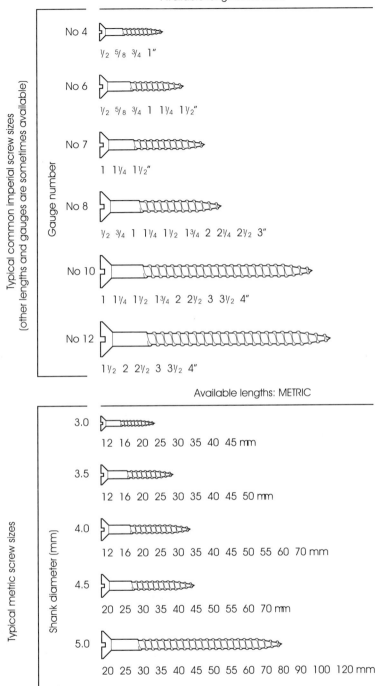

Available lengths: IMPERIAL

Typical common imperial screw sizes
(other lengths and gauges are sometimes available)

Gauge number

No 4
1/2 5/8 3/4 1"

No 6
1/2 5/8 3/4 1 1 1/4 1 1/2"

No 7
1 1 1/4 1 1/2"

No 8
1/2 3/4 1 1 1/4 1 1/2 1 3/4 2 2 1/4 2 1/2 3"

No 10
1 1 1/4 1 1/2 1 3/4 2 2 1/2 3 3 1/2 4"

No 12
1 1/2 2 2 1/2 3 3 1/2 4"

Available lengths: METRIC

Typical metric screw sizes

Shank diameter (mm)

3.0
12 16 20 25 30 35 40 45 mm

3.5
12 16 20 25 30 35 40 45 50 mm

4.0
12 16 20 25 30 35 40 45 50 55 60 70 mm

4.5
20 25 30 35 40 45 50 55 60 70 mm

5.0
20 25 30 35 40 45 50 55 60 70 80 90 100 120 mm

6.0
35 40 50 60 70 80 90 100 120 150 mm

Figure 1.72 *Standard screw sizes*

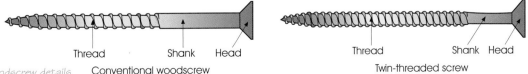

Thread   Shank   Head
Conventional woodscrew

Thread   Shank   Head
Twin-threaded screw

Figure 1.73 *Woodscrew details*

Screw heads

Both conventional and twin-threaded screws are available with a variety of screw heads (Figure 1.74).

Countersunk head screw – a flat-topped screw with a tapered bearing surface, for use where the head is required to finish flush with the surface. It fits into a countersunk hole formed into either the timber or the item of ironmongery.

Round head screw – a dome head screw with a flat bearing surface used to fix sheet material and un-countersunk ironmongery.

Raised head screw – a slightly rounded head screw with a tapered bearing surface; for use on exposed fixings in conjunction with cups or for fixing surface ironmongery such as handles and bolts.

Bugle head screws – a flat-topped screw, but with a stronger taper to the bearing surface than countersunk screws. This allows them to be driven without the need for countersinking. Mainly used as drywall screws for fixing plasterboard.

Screw slots

The recess that is cut into screw heads is designed to accommodate the tip of a screwdriver. There are varying types – the most common are illustrated in Figure 1.75.

Slotted screws – have a single groove slot machined right across the head, for a straight-tipped screwdriver.

Cross-head screws – have two intersecting slots designed to accept the tip of a matching cross-head screwdriver. The majority of screws are now of the cross-head type, which minimises slipping, especially when using power screwdrivers. In theory a different pattern cross-head screwdriver is required for different brands of screws. The two main types are Phillips and Posidrive.

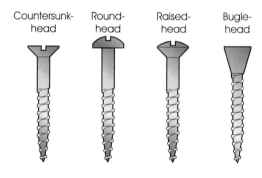

Figure 1.74 *Screw heads*

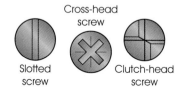

Figure 1.75 *Screw slots*

Clutch-head screws – are security, or thief-proof, screws. They can be inserted with a standard straight tipped screwdriver, but the tip rides out of the slot when turned in an anti-clockwise direction for removal.

Screw cups and covers

Many types exist, mainly for decorative purposes – the most common are illustrated in Figure 1.76.

Screw cups

These enhance the appearance of screws used for surface fixings. They also increase the bearing area and protect the countersinking from wear in situations that require screws to be removed, such as glazing beads and access panels. They are available as recessed or surface mounted, for use with countersunk or raised-head screws.

Screw covers

Used to conceal the screw head, for a neater finish while still indicating the screw's position.

Snap caps – consist of a plastic dome that snaps over the rim of a matching screw cup, after the countersunk screw has been inserted.

Cross-head cover caps – are a moulded plastic cover, which has a spigot on the underside to locate into the head of cross-head screws.

Mirror screw covers – are brass, chrome or stainless steel, often domed covers, which have a threaded spigot that screws into a threaded hole in the screw-head of special countersunk screws. They are intended to cover the head of screws used to fix mirror glass.

Using screws

The length of a screw should be approximately three times the thickness of the wood being secured, but to avoid through-penetration or dimpling the backing surface, the screw point should finish a minimum of 3 mm short (Figure 1.77).

Depending on the type of screw being used a clearance hole and countersinking should be drilled in the piece of wood being secured. A smaller pilot hole is required in the piece of wood that it is being fixed to (Figure 1.78). With smaller screws into softwood, the pilot hole may be made with a bradawl. Pilot holes for larger screws, all screws in

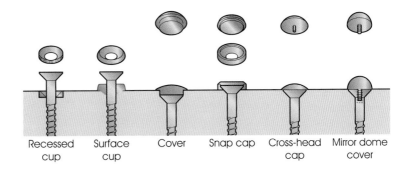

Figure 1.76 *Screw cups and covers*

Recessed cup   Surface cup   Cover   Snap cap   Cross-head cap   Mirror dome cover

Materials   Chapter 1

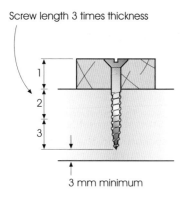

Figure 1.77 *Screw length guide*

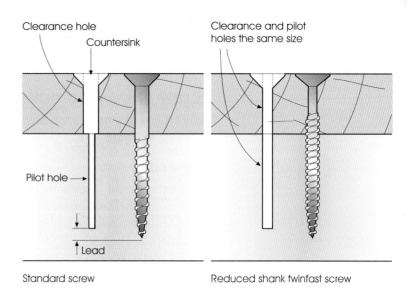

Figure 1.78 *Holes to receive screws*

hardwoods, and edge screwing into sheet materials should be drilled. The pilot hole should be shorter than the full depth of the screw. The undrilled depth is called the lead. The clearance hole for twinfast screws can be the same size as the pilot hole because they have a reduced shank gauge.

Screwdriver selection

Always use the correct size of screwdriver (Figure 1.79). This ensures that:

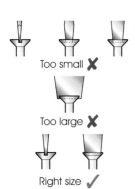

Figure 1.79 *Screwdriver selection*

- ◆ damage is not done to the screw slot;
- ◆ the screwdriver does not slip and damage the workpiece.

Slot head screws

Match the width of the blade to the slot of the screw:

- ◆ too small and it will twist or chip;
- ◆ too large and it will tend to slip out and the protruding edges will score marks in the surrounding surface as the screw is driven home.

Cross-head screws

Different blade point patterns and sizes are available to match the gauge or screw being used:

- ◆ Size 1: up to gauge 6
- ◆ Size 2: gauge 8–10
- ◆ Size 3: gauge 12–14

Nails

Nailing is the simplest way of joining pieces of wood together and securing other materials to wood. If carried out correctly, nailing can result in a strong, lasting joint. Nail types are shown in Figure 1.80.

Wire nails – Also known as French nails, are normally available in lengths from 12 mm to 150 mm. They are round in section and the larger sizes normally have a chequered head to reduce the possibility of the hammer slipping while the nail is being driven in. The top part of the shank is usually roughened to give the nail extra grip in the timber. These are general-purpose carpentry nails which can be used for all rough work where the presence of the nail head on the surface of the timber is not important, such as most carcassing, first fixing and formwork.

Oval nails – These, as the name implies, are oval in section and are available in lengths from 12 mm to 150 mm. They have a small head, which can be punched below the surface. This is an advantage when fixing timber that has to be painted because the hole is not visible when filled.

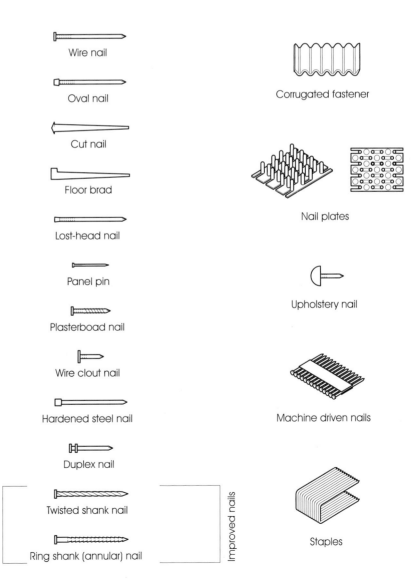

Wire nail

Oval nail

Cut nail

Floor brad

Lost-head nail

Panel pin

Plasterboad nail

Wire clout nail

Hardened steel nail

Duplex nail

Twisted shank nail

Ring shank (annular) nail

Improved nails

Corrugated fastener

Nail plates

Upholstery nail

Machine driven nails

Staples

Figure 1.80 *Typical range of nails and fasteners*

Materials Chapter 1

Cut nails – Cut from mild sheet steel, these are square in section, which gives a good grip. They are normally available up to 100 mm in length. The cut nail can be used for general construction work, but nowadays it is mainly used for fixing timber to block-work, walls, etc. They can also be used for nailing into the mortar joints of green (freshly laid) brickwork.

Floor brad – These are similar to the cut nail, but they are lighter in section. They are available in lengths up to 75 mm and are used for the surface fixing of floorboards.

Lost head nails – These are similar to the wire nail, but they have a small head, which can be easily punched below the surface of the timber without making a large hole or protrusion. Hence the name 'lost head'. They are also used for fixing timber that has to be painted and for the secret fixing of tongue and grooved flooring boards. They are available in lengths up to 75 mm.

Panel pins – A much lighter version of the lost-head nail used for fixing fine moulding around panels etc. They are normally available in lengths between 12 mm and 50 mm.

Improved nails – Wire nails with a ringed or twisted shank that can be hammered in easily, but have the holding power of a screw. They can be used for all construction work where extra holding power is required. They are available in lengths up to 75 mm.

Wire clout nails – These are a short galvanised wire nail with a large head. Their main use is for fixing roofing felt and building paper, etc. They are available in lengths between 12 mm and 25 mm.

did you know?

Hardened steel nails are also known as masonry nails.

Hardened steel nails – Similar to the lost head nail but made from a specially hardened zinc-plated steel for making hammered fixings directly into brickwork and concrete without plugging. They are available in lengths up to 100 mm.

Duplex nails – A wire nail with a double head, the lower head is driven to the surface for maximum holding power whilst leaving the upper head protruding for easy withdrawal. Mainly used for formwork and other temporary fixings.

safety tip

Goggles must always be worn when driving hardened steel nails.

Plasterboard nails – These are galvanised to prevent them rusting, and they have a jagged shank for extra grip. They are available in lengths up to 40 mm and can be used to fix plasterboard and insulating board to timber ceiling joists, studwork and battens etc.

Corrugated fasteners – Corrugated bright steel strip used to reinforce mitres and butt joints in rough framing work. The fastener is placed across the joint line and driven flush with the surface. They are available in sizes from 6 mm to 25 mm deep.

Nail plates – These are spiked or flat metal plates used for securing framework joints, normally made from zinc plated or galvanised steel. Plates are positioned or pressed into the wood. Flat plates are secured with improved nails.

Upholstery nails – Small domed head, decorative nail used for fixing upholstery fabric and trim. Normally 12 mm long and available in brass, bronze and chrome.

Machine-driven nails – Wire nails and pins packed in strips or coils for use in various types of pneumatic nailers. A wide range of types and sizes are available depending on the manufacturer, for use in structural work, right through to fine pins for fixing finishing trim.

Staples – Normally supplied in strips temporarily bonded together and used as an alternative to pneumatically driven nails, for fixing plywood, fibreboard, plasterboard, insulation board and plastic sheeting.

Using nails

Nails are driven in with an appropriate hammer (Figure 1.81). Use a Warrington or pin hammer for fine work and a claw hammer for heavier work. Always keep the hammer face clean, by rubbing it with a fine abrasive paper. A hammer's dirty face tends to slip on the nail head, damaging the workpiece and bending the nail. Keep an eye on the nail and check the angle it enters the wood. To avoid damage and bending, the hammer shaft at the moment of impact should be at right angles to the nail.

Methods of using nails are listed on the following page and illustrated in Figure 1.82.

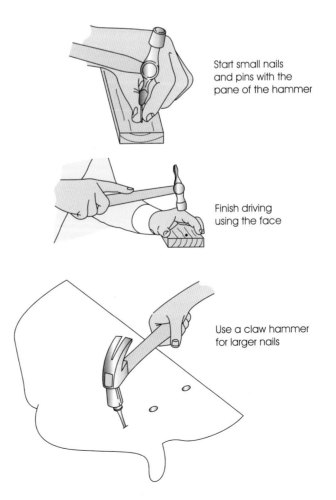

Start small nails and pins with the pane of the hammer

Finish driving using the face

Use a claw hammer for larger nails

Figure 1.81 *Driving nails*

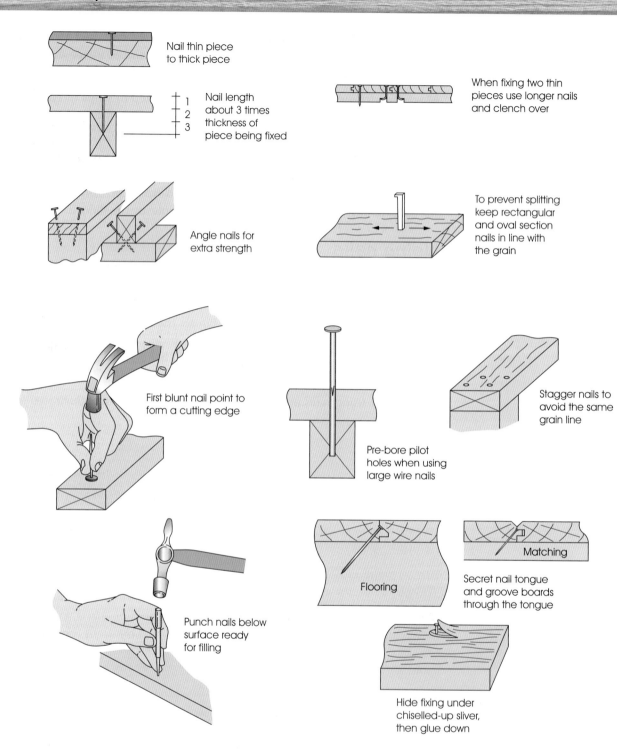

Nail thin piece to thick piece

Nail length about 3 times thickness of piece being fixed

When fixing two thin pieces use longer nails and clench over

Angle nails for extra strength

To prevent splitting keep rectangular and oval section nails in line with the grain

First blunt nail point to form a cutting edge

Pre-bore pilot holes when using large wire nails

Stagger nails to avoid the same grain line

Punch nails below surface ready for filling

Flooring

Matching

Secret nail tongue and groove boards through the tongue

Hide fixing under chiselled-up sliver, then glue down

Figure 1.82 *Using nails*

- Always nail the thin piece to the thick piece.
- Use a length of nail that is about $2\frac{1}{2}$ to 3 times the thickness of the wood it is being driven through. This gives approximately two-thirds of the nail to provide the holding power.
- When joining two thin pieces, use a nail length 4 to 6 mm longer than the combined thickness of the pieces. This allows the protruding end to be clenched over for strength.
- Where extra strength is required always dovetail or skew nail. This prevents the nails from being pulled out or working loose.

◆ Where oval or rectangular section nails are used, the widest dimension must be parallel to the grain of the timber. Their use in the opposite direction, across the grain, will normally result in the timber splitting.

◆ When nailing near the end of a piece of timber, the timber has a tendency to split. In order to overcome this, the point of the nail should be tapped with a hammer to blunt the point before the nail is used. The point of a nail tends to part the fibres of the timber and therefore split it, while the blunted end tends to tear its way through the fibres, making a large hole for itself.

◆ Pilot holes may be required or specified to receive wire nails in structural joints. These pilot holes should be up to 80% of the nail shank diameter, in order to prevent splitting.

◆ Pilot holes are also required when nailing hardwoods.

◆ Stagger nails across the grain – do not nail in the same grain line more than once, as this will split the wood. Nails in surfaces to be painted should be punched just below the surface ready for filling.

◆ Boards with tongued and grooved joints can be secret nailed through the tongue. Subsequent boards effectively hide the nail head.

◆ For general surface nailing, nail heads can be concealed by chiselling up a sliver of wood and nailing into the recess. The sliver is then glued back down to hide the nail head.

did you know?

Blunting the nail prevents splitting.

Woodworking adhesives

There is a vast range of adhesives available for use in the building industry. But it must be remembered that each adhesive has its own specific range of uses and that no one adhesive will satisfactorily bond all materials for all applications.

The main types of adhesive used in the woodworking industry are (Figure 1.83):

Animal glue – Also known as Scotch glue. Made from animal hides and bones, but rarely used now, except in some small workshops, because of the time taken to prepare and its limitations. These glues are supplied in cake form and must be broken up, soaked and heated before use.

Casein adhesive – Manufactured from soured, skimmed milk curds, which are dried and crushed into a powder. An alkali and certain fillers are added to the powder to make it soluble in water and give it its gap-filling properties. Its main use is for general joinery although it is inclined to stain some hardwoods, particularly oak. Little preparation is required as the powder is simply mixed in a non-metal container with a measured quantity of cold water and stirred until a smooth creamy consistency is achieved.

Polyvinyl acetate glue – Commonly known as PVA glue, it is a thermoplastic adhesive that is widely used for furniture and internal joinery. No preparation is required as this adhesive is usually supplied as a white creamy liquid in a nozzled, polythene bottle. It does not stain timber, but

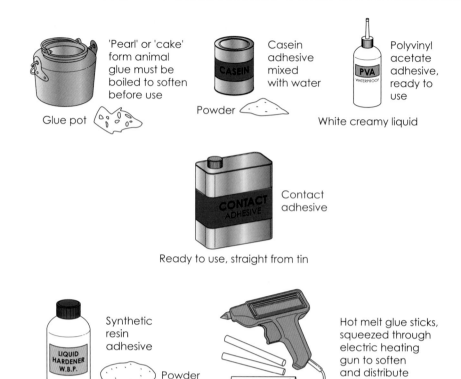

'Pearl' or 'cake' form animal glue must be boiled to soften before use

Glue pot

Casein adhesive mixed with water

Powder

Polyvinyl acetate adhesive, ready to use

White creamy liquid

Contact adhesive

Ready to use, straight from tin

Synthetic resin adhesive

Powder

Normally two-part powder and liquid

Hot melt glue sticks, squeezed through electric heating gun to soften and distribute

Figure 1.83 *Types of adhesive*

some types can affect ferrous metal. PVA glue is one glue that does not blunt the cutting edge of tools. Modified PVA adhesives are also available to increase gap filling capacity, or moisture resistance.

Contact adhesive – Available in either a solvent or water-based form. It is mainly used for bonding plastic laminates and sheet floor covering. The adhesive must be applied to both surfaces and allowed to become touch dry before being brought together. This normally takes between ten and thirty minutes, depending on the make used. Once the two surfaces touch, no further movement or adjustment is normally possible, as an immediate contact or impact bond is obtained. Modified types are available which allow a limited amount of adjustment to be made after contact. Solvent-based contact adhesives must be used in a well-ventilated area, where no smoking or naked lights are allowed. Water-based contact adhesives are safer to use but take longer to dry.

Synthetic resin adhesives

Phenol formaldehyde – Classed as a WBP (weather and boil proof) adhesive. This means that it has a very high resistance to all weather conditions, cold and boiling water, micro-organisms, dry heat and steam. It is sold in two parts, the resin and the hardener, which are mixed together as required. A disadvantage of this type is that it normally requires a very high temperature in order to set. It is mainly used for exterior plywood, exterior joinery and timber engineering.

Resorcinol formaldehyde – is completely water resistant and, like phenol formaldehyde, is classed as a WBP adhesive. It is sold as a liquid to which a powder or liquid hardener is added. It is mainly used for timber engineering and marine work.

Urea formaldehyde – is classed as an MR (moisture resistant) adhesive. This means that it is moderately weather resistant and will withstand prolonged exposure to cold water, but very little to hot water. It is also resistant to micro-organisms. If urea formaldehyde is fortified by including resorcinol or melamine in the hardener, it can be classed as a BR (boil resistant) adhesive. This means that it has a good resistance to boiling water and fairly good weather resisting properties. It is also resistant to cold water and micro-organisms. Urea formaldehyde is available as either a one-part adhesive that is mixed with water or as a two-part adhesive with a separate hardener. Its main uses are for general joinery, furniture and plywood.

Hot melts – These are made from ethylene vinyl acetate and are obtained in a solid form. They become molten at very high temperatures and set immediately on cooling. Small hand-held electric glue guns and automatic edging machines normally use this type of adhesive. Hot melt glue is also available in thin sheets for veneering work. The glue is placed between the veneer and board, then activated with a heated domestic iron.

Gap filling – Refers to adhesives that are capable of filling gaps up to 1.3 mm wide without affecting the strength of the joint.

safety tip

Always follow the safety procedures when using adhesives

Storage life

Refers to the period in which the materials forming the glue can be stored and still remain usable. After being stored for a certain length of time, glue becomes useless. This time varies with different types of glue. All glues should be stored in cool frost-free conditions.

Pot life

Refers to the time available for using the glue after it has been mixed. After a certain amount of time most glues become too thick to work with ease.

Pot life and setting time depends to a large extent on the temperature.

Adhesive safety

Adhesives can be harmful to your health. Many adhesives have an irritant effect on contact with the skin and may result in dermatitis. Some are poisonous if swallowed, while others can result in narcosis if the vapour or powder is inhaled. These and other harmful effects can be avoided if proper precautions are taken:

- ◆ Always follow the manufacturer's instructions.
- ◆ Always use a barrier cream or disposable protective gloves.
- ◆ Do not use those with a flammable vapour near a source of ignition.
- ◆ Always provide adequate ventilation.
- ◆ Avoid inhaling toxic fumes or powders.
- ◆ Thoroughly wash your hands, before eating or smoking and after work, with soap and water or an appropriate hand cleanser.
- ◆ In the case of accidental inhalation, swallowing or contact with eyes, medical advice should be sought immediately.

Materials

Chapter 1

Using adhesives

There are various guidelines which are important to follow when applying adhesives (Figure 1.84). Always read the manufacturer's instructions because general rules may not always apply. Here are some practical tips.

Timber preparation

The timber should be seasoned, preferably to the equilibrium moisture content it will obtain in use. The timber should be planed to a smooth even surface and all dust must be removed before gluing. The gluability of a timber surface deteriorates with exposure; therefore the time between preparation and gluing should be as short as possible.

Adhesive preparation

Adhesives that consist of two or more components must be mixed accurately in accordance with the manufacturers instructions. It is normally advisable to prepare the component parts by weight rather than volume.

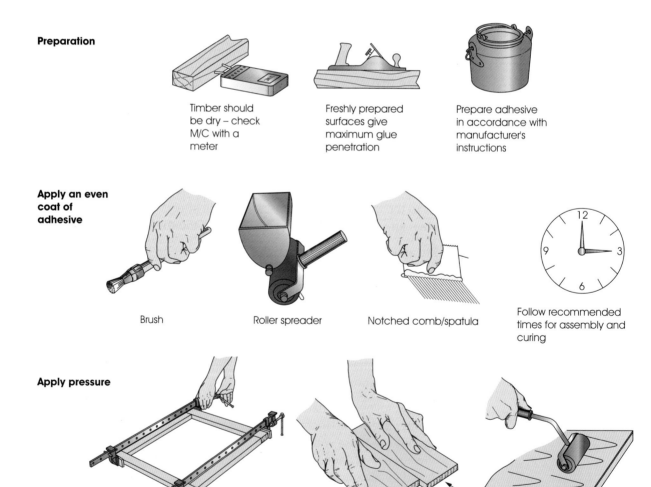

Preparation

Timber should be dry – check M/C with a meter

Freshly prepared surfaces give maximum glue penetration

Prepare adhesive in accordance with manufacturer's instructions

Apply an even coat of adhesive

Brush

Roller spreader

Notched comb/spatula

Follow recommended times for assembly and curing

Apply pressure

Cramp-up frames

Edge joints may be rubbed to and fro

Surface laminates and veneers may be rolled

Figure 1.84 *Using adhesives*

Chapter 1 Materials

Application of adhesives

They may be applied either by brush, roller, spray, spatula or mechanical spreader. For maximum strength a uniform thickness of adhesive should be applied to both sides of the joint. Adhesives that set by chemical reaction begin to react as soon as the components are mixed. This reaction rate is dependent mainly on the temperatures of the adhesive, the timber and the surrounding room or workshop. These factors must be taken into account to ensure that pot life of the adhesive is not exceeded; otherwise the strength of the joint will be affected.

Assembly time

This is the elapsed time between the application of the adhesive and the application of pressure. Some adhesives benefit in strength if they are allowed to partly set before the surfaces are brought into contact (open assembly time). Other adhesives require a period to thicken while the surfaces are in contact but before pressure is applied (closed assembly time). These times will be specified by the manufacturer and must be carefully controlled to ensure that the adhesive is squeezed out when the pressure is applied.

Pressure

Pressure should be applied to the glued joint in order to:

- ◆ spread the adhesive uniformly;
- ◆ squeeze out excess adhesive and pockets of air;
- ◆ ensure close contact between the two adjoining surfaces.

This pressure must be sustained until the joint has developed sufficient strength (cramping period). The application of heat will speed the development of strength and therefore reduce the cramping period.

Curing

This is the process that leads to the development of full strength and resistance to moisture. It starts during the cramping period and is completed while the components are in storage prior to use (conditioning period). Curing is also dependent on temperature and can be speeded up by heating.

activity

The following timber and board materials are required in order to make 10 door linings and the panels in the bottom of 10 brought-in timber framed doors.

- ◆ 20 lengths of European redwood, finished size $38 \times 140 \times 2050$ mm.
- ◆ 10 lengths of European redwood, finished size $38 \times 140 \times 800$ mm.
- ◆ 10 pieces of 12 mm birch-faced plywood (to be clear finished on both faces) 750×650 mm, to be cut from 2440×1220 mm sheets.

Determine the total length of timber including a 15% cutting allowance and the number of plywood sheets required for the job. Then write or type a fax to request the price of the materials required from a local supplier.

1. Name the term used to describe the usable amount of time an adhesive has after it is mixed ready for use.

2. State the reason why nail points may be blunted before use.

3. Name an adhesive suitable for internal joinery or cabinet construction.

4. A sheet of plywood is 1220mm × 2440mm A/bal WBP. Explain what this means

5. Produce a sketch to show the following screw terminology:
   a) countersink
   b) clearance hole
   c) pilot hole
   d) lead

6. Explain the essential safety precaution to be taken when using a contact adhesive.

7. Explain the difference between MDF and MRMDF.

8. Explain what is meant by 'improved' nail.

9. Produce a sketch to show the orientation of plies in a sheet of 6-ply.

10. A specification indicates that MFC is to be used. State what this means.

Handtools

This chapter is intended to provide the reader **with an overview of the** types, use, maintenance and sharpening of woodworking handtools. Although its content is not assessed directly, knowledge of its contents is assumed and assessed in Wood Occupations units VR 05, VR 06, VR 07, and VR 08 at Level 1. Knowledge is also assessed in both Site Carpentry and Bench Joinery at Level 2.

In this chapter you will cover the following range of topics:

- Measuring and marking out
- Squares and sliding bevels
- Gauges
- Saws
- Planes
- Chisels and gouges
- Drills and braces
- Fixing tools
- Finishing tools and abrasives
- Miscellaneous tools
- Workshop equipment
- Site equipment

What's required in VR 05, VR 06, VR 07 & VR 08?

To successfully complete these units you will be required to demonstrate your skill and knowledge of the following processes:

- Using and maintaining woodworking handtools.

You will be required practically to:

- Use a range of hand tools for
    - measuring and marking;
    - ripping and crosscutting;
    - planing, rebating and grooving;
    - drilling, boring and recessing;
    - driving fixings and inserting screws.
- Maintain and sharpen woodworking handtools.
- Observe safe working practices and follow instructions when using, maintaining and sharpening woodworking handtools.

The skills required to use and sharpen handtools are relatively easy to acquire. However they take much patience and practice to perfect, but the rewards will be with you for life. Tool skills, coupled with woodworking practices, will be evident in the quality of the finished product, for better or worse!

Measuring and marking out

Accuracy is a simple matter: take care to mark the timber to size and then cut to the marked line. This way your components cannot fail to fit exactly together.

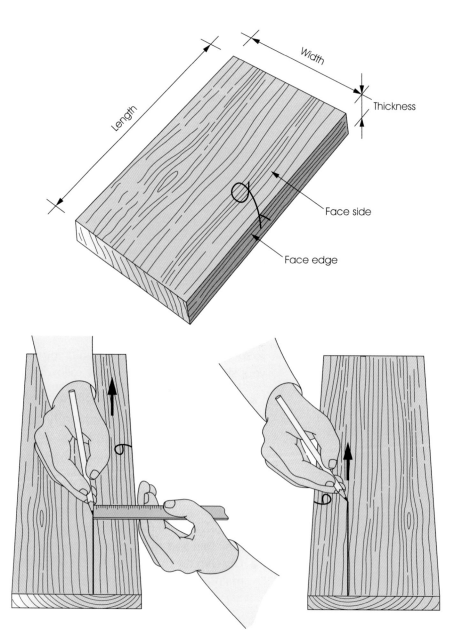

Figure 2.1 *Marking out and ruling parallel lines*

Timber cannot be accurately marked out for cutting until it has been first prepared with flat faces and square parallel edges. Face and edge marks are applied to indicate that the two adjacent surfaces are true and square to each other (Figure 2.1). It's from these surfaces that marking out progresses. The face side mark is a looped character, like a figure '9', with its leg extending to the face edge. The face edge mark is indicated by an upside down letter 'V' with its apex pointing towards the face side and joining up with the leg of the face side mark.

Rules and tape measures

You should be aware of the trade terms used in measuring a piece of wood.

The length of a piece is taken along the grain. The width of a piece is the distance across the widest face from edge to edge. The thickness of a piece is the distance between the faces.

Rules

Various types of rule are used to mark out and check dimensions.

Four-fold metre rule – was traditionally the basic woodworkers' rule (Figure 2.2). They are still available in either wood or plastic and with markings in imperial or metric measurements, or both. Some patterns have a chamfered edge of the first fold for more accurate marking off.

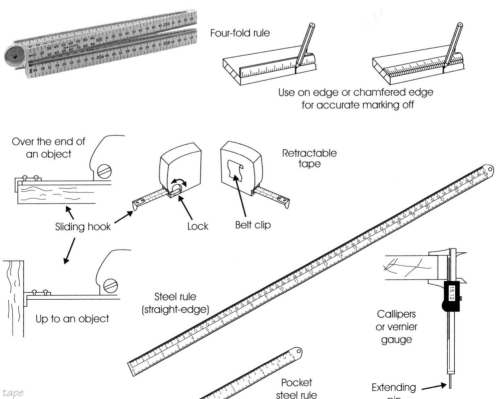

Four-fold rule

Use on edge or chamfered edge for accurate marking off

Over the end of an object

Sliding hook

Up to an object

Lock

Belt clip

Retractable tape

Steel rule (straight-edge)

Pocket steel rule

Callipers or vernier gauge

Extending pin

Figure 2.2 *Rules and tape measures*

Handtools Chapter 2

did you know?

All marking out should be done from either the face side or face edge.

did you know?

Steel rules should be smeared with a light oil to prevent them from rusting.

safety tip

Take care with sharp corners or irregular grain to avoid cut fingers and splinters.

Retractable tape measures – are used by many woodworkers: they fit easily in a pocket or clip over your belt. The flexible steel tape is enclosed in a compact spring-loaded container. Many models can be locked in an extended position and be retracted back into its case automatically. Typical lengths for woodworking use are between 3 and 8 metres. A hook at the end of the tape aids measuring long lengths. It is designed to slide back and forth, thus compensating for its thickness when measuring up to an object or over the end of an object. For accuracy, check before every use that the slide is working and not damaged or stuck with surplus glue.

Steel rules – are considered essential for accurate workshop use: a 600 mm or 1 m long rule by the bench and a 150 mm rule for your pocket. The longer steel rules also have a dual use as a straight-edge.

Dial callipers or vernier gauge

These are used in the workshop principally for accurate thickness measurement. The least expensive types, with either a dial scale or digital readout, and capable of measuring to 0.25 mm, are the most suitable for woodwork.

Other uses for rules

Rules are used to divide the width of a board into equal parts (Figure 2.3):

◆ Place the rule diagonally across the board.
◆ Adjust the rule to give the required number of parts.
◆ Mark off with a pencil.

Rules can be used as an aid when 'ruling off' lines parallel to an edge, for saw cuts, etc. (Figure 2.1.):

◆ Position the pencil against the end of the rule.
◆ Grip the rule between the thumb and first finger, using the knuckle as a guide against the edge of the wood, while pulling the rule towards you.

When marking parallel lines close to an edge dispense with the rule and use the tip of the first finger as a guide.

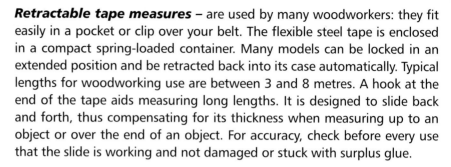

Example to divide board into 7 equal parts: angle rule across board to read a whole number which is easily divided by 7, say 350 mm, mark off at 50 mm intervals

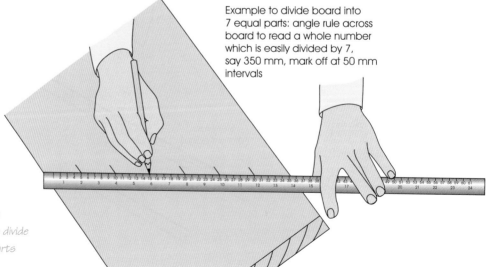

Figure 2.3 *Use of a rule diagonally to divide into equal parts*

Squares and sliding bevels

Try squares

These are used to mark lines at right angles across from the face side or face edge of a piece of wood (Figure 2.4). In addition, as the name suggests, they are used to test pieces of wood or made-up joints for square.

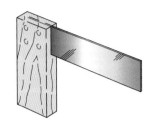

Figure 2.4 *Try square*

The woodworker's try square has a steel blade which is often 'blued' to prevent it rusting and a rosewood stock. In use the blade of the square should rest flat on the piece with the stock pressed up firmly to the edge. Mark out using the outer edge of the blade, keeping the inner edge for testing (Figure 2.5).

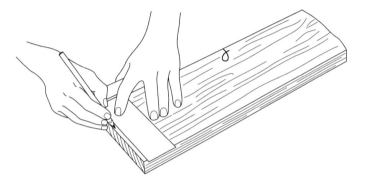

Figure 2.5 *Using a try square*

Most marking out is done with a pencil; however shoulder lines for joints are often marked with a knife to cut the surface fibres cleanly, leaving a guide for the saw to follow.

Checking for accuracy – To ensure accuracy you should periodically test your square for trueness:

- Place the stock against an edge known to be perfectly straight and mark a line (Figure 2.6).
- Turn the square over and check the previously marked line.
- If the square is 'true', they line up perfectly.
- Any misalignment between the blade and line will result in inaccuracies. Take time to replace the square, taking extra care next time not to misuse it or accidentally drop it.

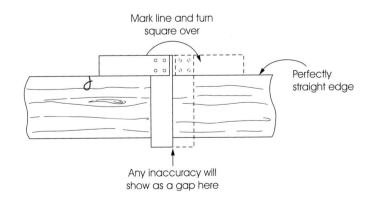

Mark line and turn square over

Perfectly straight edge

Any inaccuracy will show as a gap here

Figure 2.6 *Checking a square for trueness*

Checking for square – When using a square to check the squareness or flatness of a piece of timber, a good light source behind the blade is essential:

◆ Place the stock against the face of the piece with the blade just touching the edge; slide the square along the edge, watching out for any light appearing under the blade (Figure 2.7).
◆ Flatness is checked with the blade placed on its edge across the face. Again slide the square along the piece – any unevenness is shown up by light appearing under the blade.

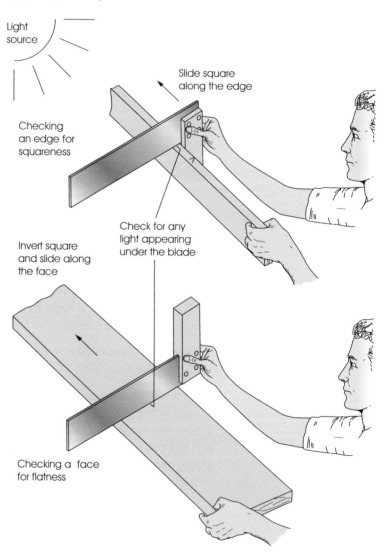

Figure 2.7 *Using a try square to check for squareness and flatness*

Mitre square

A mitre square is similar to a try square with a steel blade and a rosewood stock, but the blade is set permanently at an angle of 45° to the stock, for marking out and testing mitres (Figure 2.8).

Combination square

This is an metal construction with a stock that can slide along the blade (Figure 2.9). It is useful for ruling off and measuring depths as well as

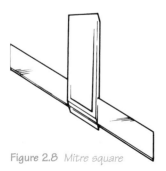

Figure 2.8 *Mitre square*

marking out and testing both right angles and mitres. In addition the blade can be removed for use as a rule. Often the stock incorporates a spirit level for checking plumb and level. Before use, ensure that the screw securing the blade to the stock is tight. See Figure 2.10 for uses.

Figure 2.9 *Combination square*

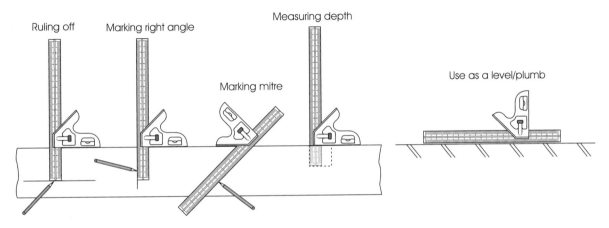

Figure 2.10 *Use of a combination square*

Sliding bevel

As an adjustable square this can be set to any angle and is used for marking out and testing angles, bevels and chamfers (Figure 2.11). It consists of a rosewood stock and a sliding steel blade, which is normally secured with a screw or wing nut.

Figure 2.11 *Sliding bevel*

Using marking knives with rules and try squares

These are used in preference to a pencil for accurately marking shoulder lines when using rules and try squares (Figures 2.12 and 2.13). This procedure is used before cutting with a tenon saw, particularly when using hardwood. The knife is sharpened like a chisel with one bevel. The flat edge is used against the blade of a try square and the bevel on the waste side of the joint.

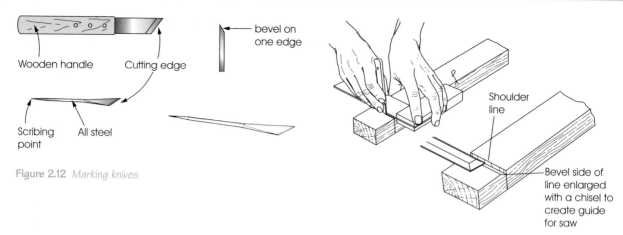

Figure 2.12 *Marking knives*

Figure 2.13 *Use of a marking knife*

did you know?

With the knife held at 90° to the job, draw your knife along the outside edge of the square to give a V-line across the piece. This cleanly cuts the surface fibres to prevent tearing. Before sawing, the bevel side of the line may be enlarged with a chisel, to create a channel that acts as a guide for the saw, and to aid a clean start.

Gauges

The surest way to score cutting lines on a piece of timber, parallel to an edge or end, is with a gauge. They consist of a wooden stem containing the steel marking pin or blade and a wooden stock, which slides along the stem and is locked off in the required position with a turn-screw (Figure 2.14). Gauges are usually made from either beech or rosewood. Better quality gauges have brass strips inlaid across the working face of the stock to reduce wear.

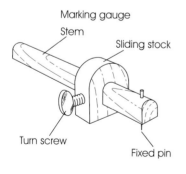

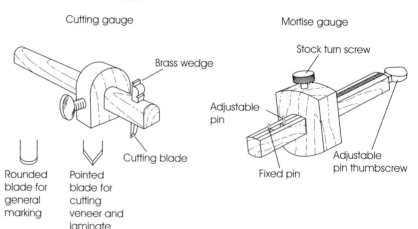

Figure 2.14 *Carpenter's gauges*

Marking gauge – This is fitted with a steel pin or spur for marking with the grain or across the end grain. Since the pin tends to tear the surface fibres it is not recommended for use across the grain.

Cutting gauge – is fitted with a blade in place of the pin, to cut a clean line across the face grain. It is also useful for cutting laminate strips, thin wood sections and easing the sides of grooves or rebates.

Mortise gauge – has two pins, with the inner one adjustable to enable the scoring of parallel lines for mortise and tenon joints.

Combination gauge – is dual purpose, and can perform as either a marking gauge or a mortise gauge by simply turning it over.

Using gauges

The correct grip is essential to scoring accurate gauge lines. Grip the stock and stem, placing your thumb on the stem close to the pin and your index finger over the top of the stock to control the gauge tilt (Figure 2.15). Your other fingers hold the stem and push the stock against the workpiece to be marked. Tilt the gauge with the pin trailing into the work and steadily slide the gauge forward. If the point digs into the work, push down on your index finger to adjust the tilt. To centre the gauge's pin on your workpiece for halving joints, etc:

◆ Set the stock using a rule.
◆ Mark from both sides.
◆ Reset the stock so that the pin is central between the two marks.
◆ Check that it is central, again from both faces.

Fine adjustment of the gauge can be made by gently tapping the stem on the bench top. Ensure the screw that secures the stock to the stem is tight before use.

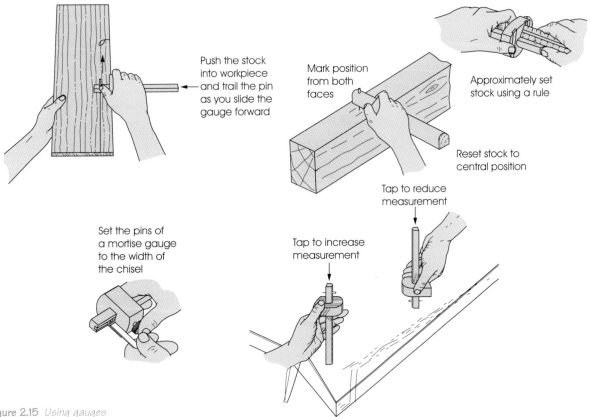

Push the stock into workpiece and trail the pin as you slide the gauge forward

Mark position from both faces

Approximately set stock using a rule

Reset stock to central position

Tap to reduce measurement

Set the pins of a mortise gauge to the width of the chisel

Tap to increase measurement

Figure 2.15 *Using gauges*

Handtools Chapter 2

To set a mortise gauge, slacken the screw that secures the stock to the stem. Adjust the thumbscrew, so that the distance between the pins is the exact width of the chisel. Set the stock so that the pins are in the required position; check from either face for central mortises, or line up the mortise with the rebate or mould for offset mortises. Re-tighten the stock, securing the screw before use.

Cutting gauges are set up and used in the same way as a standard marking gauge. Like a marking knife, its blade should be sharpened on one side only with the bevel facing the stock.

Saws

There is a wide variety of saws available to the woodworker. In general they can be considered in four main groups:

◆ handsaws
◆ backsaws
◆ frame saws
◆ narrow-blade saws.

Cutting action

All saw teeth are 'set' or bent outwards to make a saw cut or 'kerf' (Figure 2.16). This gives the blade clearance and prevents it jamming or binding in the timber.

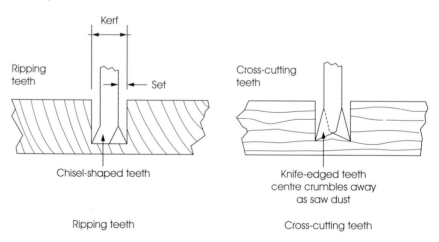

Figure 2.16 *Saw teeth*

Ripping teeth – are chisel-end teeth and the cutting action is in fact like a series of tiny chisels, each cutting one behind the other.

Cross-cutting teeth – have knife points to sever the fibres of the timber. These points are so arranged that they cut two knife lines close together. The centre fibres between these knife lines crumble away as sawdust.

In order to prevent the saw blade from jamming in the timber, its teeth are 'set', that is each alternate tooth is bent outwards to make a saw-cut, or kerf, which is just wide enough to clear the blade.

did you know?

The width of a saw-cut is known as its kerf.

Chapter 2 Handtools

Handsaws

These are used for the preliminary cutting of components to size and take three forms (Figure 2.17):

◆ The ripsaw is the largest, up to 750 mm long with 3–6 teeth per 25 mm which is 3–6 teeth per inch (TPI) in imperial measurements. They are used for cutting along the grain only.
◆ The cross-cut saw is up to 650 mm long with 6–8 teeth per 25 mm or TPI and is mostly used for general purpose cross-cutting to length.
◆ The panel saw is the smallest of the handsaws at around 550 mm in length, with 10 teeth per 25 mm or TPI. Generally used for cutting up sheet material.

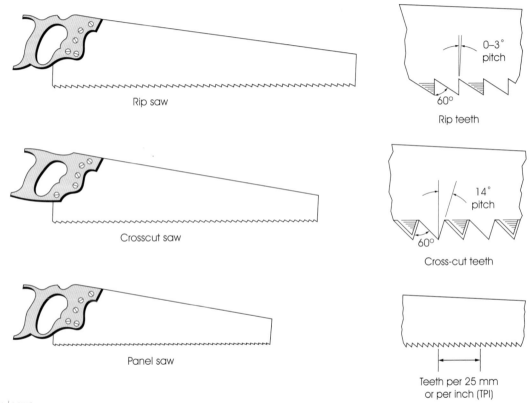

Figure 2.17 *Handsaws*

Cross-cut and panel saws – can also be used for ripping but progress is slower.

Rip saws – are almost impossible to use for cross-cutting as the tooth shape causes it to 'jar' and tear the fibres.

Good quality handsaws have a tapered ground blade that is thicker near the teeth and gets thinner towards its upper edge. This gives it added clearance in the kerf and aids to guide the saw.

Saw teeth will dull (become blunt) very quickly when cutting MDF and chipboard. Many woodworkers will purchase an inexpensive hard-point saw for such occasions. These have teeth that retain their edge longer but when they do dull they have to be thrown away as they cannot be re-sharpened.

did you know?

Check the metal quality when purchasing a saw: a firm tap with your knuckle should produce a nice ringing sound, not a flat thud. Try bending the blade to a 'U' shape – on release, it should return perfectly straight.

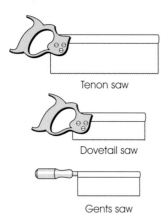

Tenon saw

Dovetail saw

Gents saw

Figure 2.18 *Backsaws*

did you know?

Use a tenon saw with its teeth re-cut square across the face for ripping fine joints, rather than a rip saw.

Backsaws

These are used for benchwork. They have their upper edge stiffened with a brass or steel folded strip, to prevent twisting or buckling during use (Figure 2.18).

Tenon saw – used for cutting shoulders of joints and general bench cross-cutting. Blade length ranges from 250 mm to 350 mm with 10–14 teeth per 25 mm (TPI). Some woodworkers find it useful to have two tenon saws, one with bevelled teeth for cross-cutting and the other with teeth re-cut square across their face for cutting along the grain.

Dovetail saw – used for cutting dovetails, mouldings and other delicate work. Blade lengths range from 200 to 250 mm with 16–20 teeth per 25 mm (TPI).

Gents saw – for the finest cutting of mouldings, etc. Blade lengths range from 100 to 250 mm with up to 32 teeth per 25 mm (TPI).

Frame saws

In general these have a fine replaceable blade which is held in tension by a frame (Figure 2.19).

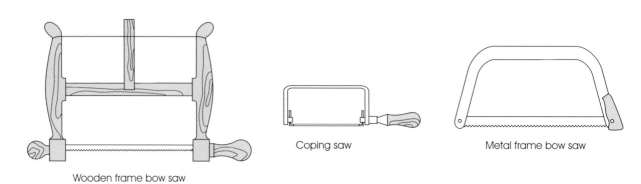

Wooden frame bow saw

Coping saw

Metal frame bow saw

Figure 2.19 *Frame saws*

Bow saw – the traditional all-wooden version was once used as a general bench saw. Its blade could rotate in the frame and be used for ripping, cross-cutting and cutting curves. The modern metal frame saw with the same name is really intended for cross-cutting carcassing timber (floor joists, etc.) especially in damp timber as the narrow blade gives less resistance than a handsaw.

Coping saw – designed to make curved cuts in wood and board material. A fret saw is similar, but has a deeper bowed frame for cutting further from the edge of a piece.

Narrow-blade saws

Keyhole or padsaw – mainly used for cutting keyholes and other small shapes and holes away from the edge of the timber (Figure 2.20).

Compass saw – mainly used for cutting larger shapes or holes away from the edge of the material.

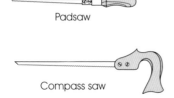

Padsaw

Compass saw

Figure 2.20 *Narrow-blade saws*

Other types of saw

Japanese saws – are becoming popular for fine woodwork: they have long teeth, a thin blade, fine kerf and generally cut on the pulling rather than the pushing stroke (Figure 2.21).

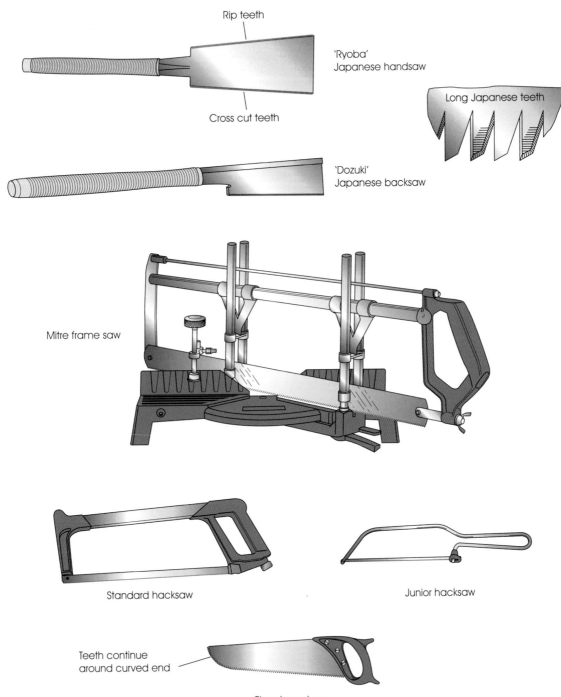

Rip teeth

'Ryoba'
Japanese handsaw

Long Japanese teeth

Cross cut teeth

'Dozuki'
Japanese backsaw

Mitre frame saw

Standard hacksaw

Junior hacksaw

Teeth continue
around curved end

Floor board saw

Figure 2.21 *Other types of saw*

Mitre saws – a modern frame saw set in a metal jig that can be set to cut at predetermined angles. Often used for the mitring of picture framing and fine mouldings, also for the accurate general cross-cutting of small components.

Hack saws – Frame saws used by the woodworkers to cut metal sections.

Floorboard saw – the front end of this saw is curved with teeth continuing round the curve. It can be used to start a cut in the middle of the work. Its main use is for cutting heading joints, when forming access points in existing timber-boarded floors.

Method of saw use

This will vary depending on the material being cut, the direction of cut and where the work is being undertaken (Figures 2.22 to 2.27).

♦ Grip the saw handle firmly: It helps to extend your index finger along the side of the handle as a guide to give you maximum control as you saw.

♦ Starting a cut: Position your thumbnail on the marked line to give your saw a starting location. Draw the saw backwards a few times, starting with short strokes and a low angle. Slacken your grip and continue sawing using full blade-length strokes.

♦ Let the saw do the work: Use a light pressure in the forward stroke, relaxing on the return. Your concentration should be on 'thinking' the saw along the line.

♦ Ripping on stools: Use a cutting angle of about 60° (Figure 2.23). Two saw stools may be required for long lengths or wide boards. Start from beyond the first stool, continue between the stools and complete the cut beyond the second stool. If the grain closes on the blade causing binding, insert a wedge in the kerf to keep it open and apply a little candle wax to lubricate the blade.

♦ Cross-cutting on stools: Use a cutting angle of about 45° (Figure 2.24). Clamp the board secure with the aid of your knee. Support the overhanging end and use very gentle strokes towards the end of the cut to avoid splitting.

♦ Ripping in the vice: Place the piece to be cut vertically in the vice when cutting the full length (Figure 2.25). Reverse the piece in the vice and saw from the other end to complete the cut. When ripping joints, angle the piece in the vice, so that both the ripping line and the line across the end grain can be seen. The aim when ripping down a pencil line is to saw on the waste side, just leaving the pencil line visible on the job. With gauged lines you should saw a 'half mark'. Again on the waste side, take out half the gauged line and leaving a trace of it still present.

♦ Cross-cutting on the bench: Use a bench hook held in the vice or butting over the front of the bench (Figure 2.26). This allows you to secure your job and also protect the bench.

♦ Coping saws: Often used to cut out the waste when forming dovetail joints. This is best achieved with the piece held vertically in the vice and about 50 mm protruding above the bench surface. When using coping or padsaws to cut internal shapes, pre-drill a hole on the waste side, to give a starting point (Figure 2.27).

did you know?

Always draw a saw backwards (towards you) on the first few strokes to give the saw a start.

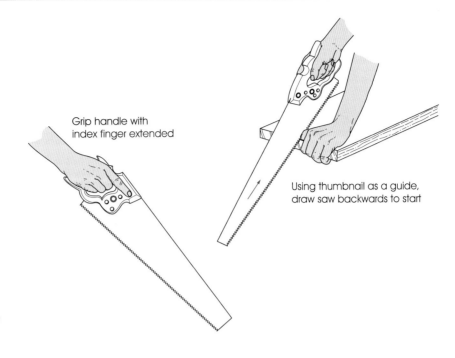

Grip handle with
index finger extended

Using thumbnail as a guide,
draw saw backwards to start

Figure 2.22 *Method of saw use*

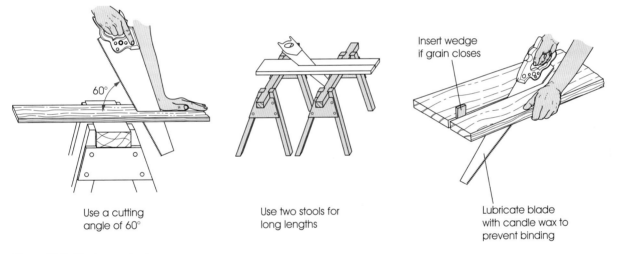

60°

Use a cutting
angle of 60°

Use two stools for
long lengths

Insert wedge
if grain closes

Lubricate blade
with candle wax to
prevent binding

Figure 2.23 *Ripping on stools*

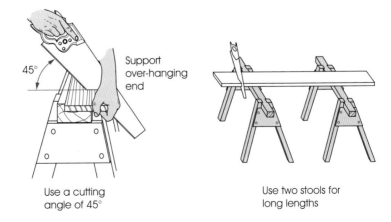

45°

Support
over-hanging
end

Use a cutting
angle of 45°

Use two stools for
long lengths

Figure 2.24 *Cross-cutting on
stools*

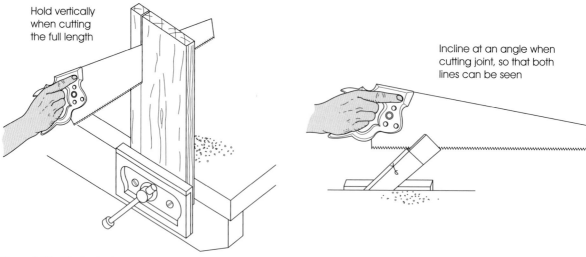

Hold vertically when cutting the full length

Incline at an angle when cutting joint, so that both lines can be seen

Figure 2.25 *Ripping in the vice*

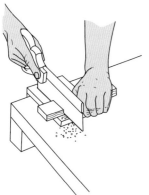

Figure 2.26 *Cross-cutting on the bench*

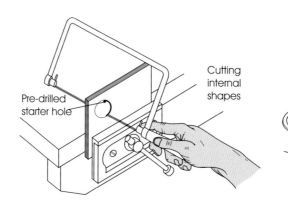

Pre-drilled starter hole

Cutting internal shapes

Removing waste

Figure 2.27 *Use of coping saw*

did you know?

It is preferable to have a second handsaw kept exclusively for cutting sheet materials, as the resin/ adhesive they contain will quickly blunt the teeth.

Cutting sheet material

Use a fine tooth panel saw when cutting manufactured boards. Lay battens across two saw stools for support and place the sheet on top (Figure 2.28). To prevent 'break-out' when cutting across the grain of plywood or when cutting faced or laminated boards, pre-score both faces of the board with a marking knife and cut on the waste side. Alternatively, apply masking tape to both sides of the board and cut through it.

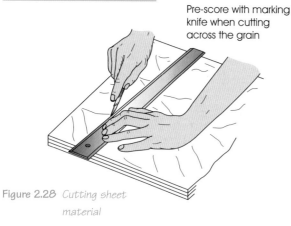

Pre-score with marking knife when cutting across the grain

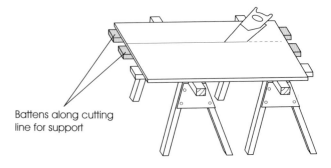

Battens along cutting line for support

Figure 2.28 *Cutting sheet material*

Saw care and maintenance

◆ Protect saw teeth when not in use with a plastic or timber sheath.
◆ Hang up or store flat in a box to prevent distortion (Figure 2.29).
◆ Lightly smear the blade with oil to prevent rusting and pitting.
◆ Avoid cutting reclaimed timber as it often contains hidden nails and screws.

did you know?

A dull saw will show 'shiners' on the tooth tips.

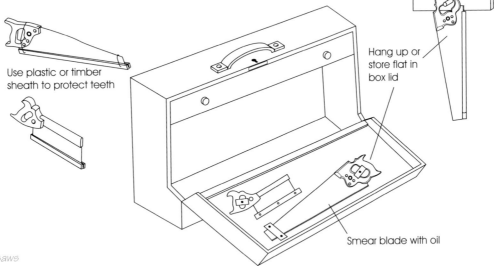

Use plastic or timber sheath to protect teeth

Hang up or store flat in box lid

Smear blade with oil

Figure 2.29 *Care of saws*

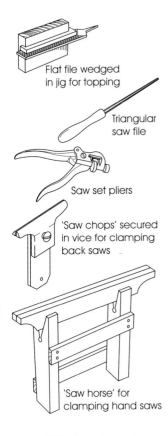

Flat file wedged in jig for topping

Triangular saw file

Saw set pliers

'Saw chops' secured in vice for clamping back saws

'Saw horse' for clamping hand saws

Figure 2.30 *Saw sharpening equipment*

To keep saws working efficiently they must be regularly sharpened and set. Signs of a dull (blunt) saw are 'shiny' teeth tips, increased effort required during sawing and the saw binding in the cut. The techniques used and equipment required for saw maintenance are similar for all types of saw. A saw in poor condition, due to inaccurate sharpening or misuse (cutting through nails or screw, etc.) may require all of the following operations:

◆ topping
◆ shaping
◆ setting
◆ sharpening.

The equipment required (Figure 2.30) includes the following:

◆ Flat file – for topping
◆ Triangular saw file – must be the right size for the saw to be sharpened. In general the saw tooth should rise just over halfway across the file face.
◆ Saw set pliers – for setting
◆ Saw clamp – a means of clamping the saw blade firmly along the entire length, to prevent vibration during filing. Saw 'chops' – for back saws and a saw 'horse' for hand saws. These are simple wooden devices easily made by the woodworker.
◆ Oilstone or slip-stone – for side dressing (medium grade)

Handtools Chapter 2

Topping

This is the first stage of sharpening a saw when the teeth are at different heights due to poor sharpening in the past (known as cows and calves) or the saw has worn unevenly, possibly by contact with metal objects (Figure 2.31). To hold the file square to the saw blade and protect your fingers, wedge the file in a grooved block of timber. Lightly draw the file along the entire length of the saw. Remove just enough metal from the teeth tips to bring the larger teeth down to the same height as the smaller ones (levelling up the cows and calves) and to produce a 'shiner' (newly filed surface) on the tip of every tooth.

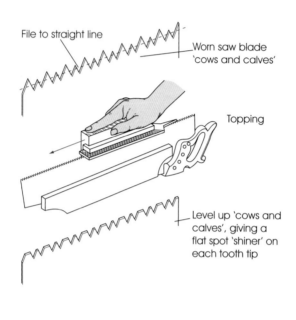

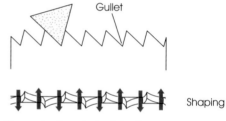

File each alternate tooth to shape and remove shiner, reverse saw and repeat to shape other teeth

File ripsaws at 90° until shiner disappears

File crosscut saws at 60°

Setting

Squeeze pliers on each alternate tooth, reverse saw and repeat to set other teeth

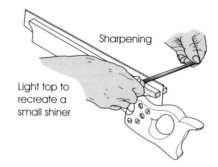

Sharpening

Light top to recreate a small shiner

File bevel on each alternative tooth, reverse saw and repeat to sharpen other teeth

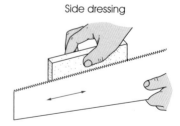

Side dressing

Lightly draw oil stone up and down saw on both sides to remove burrs

Figure 2.31 *Saw maintenance*

Shaping or reshaping

This is the filing of the teeth to their original size and shape. With the blade securely clamped, start filing from the handle end on the front edge of the first tooth bent away from you. Press the file into the gullet and file horizontally across at right-angles to the blade. Two or three firm strokes may be required to remove half the 'shiner'. Repeat the process on each alternative gullet up to the tip of the blade. Turn the saw around and repeat the process on the other teeth not yet filed. This time removing the other half of the shiners, leaving all the teeth the same size and shape.

Setting

This is the bending over of the teeth tips to give the blade clearance in the kerf. Adjust the saw set pliers for the correct number of teeth (TPI). Place the set over the blade, squeezing the handles firmly to set each alternate tooth. Turn the saw around and repeat the process to set the remaining teeth.

Sharpening

This puts a cutting edge on the teeth. Firstly, lightly draw a flat file along the blade to recreate small shiners on the top of each tooth. Working from one end with the triangular file held horizontal, and at the correct angle for the type of saw, lightly file each alternate tooth until its shiner disappears. Filing the bevel on the front edge of one tooth also puts the bevel on the back edge of its adjacent tooth. After reaching the end, turn the saw around and repeat the process. The final stage of sharpening is side dressing, to remove burrs and ensure all the teeth are evenly set. Lightly draw a medium grade oilstone or slip-stone up and down the teeth on either side of the saw blade.

Although it is possible to maintain saws yourself, many woodworkers will send their saws periodically to a 'saw doctor' (specialist in saw maintenance) for a full reshape and sharpening, then extending the cutting life themselves by 'touching up' with a set and sharpen.

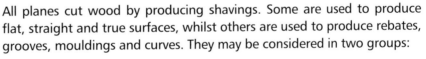

Planes

All planes cut wood by producing shavings. Some are used to produce flat, straight and true surfaces, whilst others are used to produce rebates, grooves, mouldings and curves. They may be considered in two groups:

◆ bench planes (Figures 2.32 to 2.40);
◆ specialist planes (Figures 2.41 to 2.43).

Traditional bench planes were made of wood. Some are still available but are difficult to adjust.

did you know?

Use your longest plane to true long edges.

Bench planes

Various sizes are used by most woodworkers (Figure 2.32) to dimension timber, prepare surfaces prior to final finishing, and to trim and fit both joints and components.

Smoothing plane

Jack plane

Try plane

Figure 2.32 *Bench plane types*

Smoothing plane – is a finishing plane. It is used for smoothing up a job after the jackplane has been used, and for general cleaning up work. Its length is 250 mm.

Jack plane – is mainly used for reducing timber to the required size and for all rough planing work. Its length of 375 mm enables it to be used satisfactorily for straight planing. Being an all-round, general-purpose plane it is ideal for both site and bench work.

Try plane – is the largest plane a craftsman uses. Its main use is for straight planing and levelling. This type of plane varies from 450 mm to 600 mm in length.

All bench planes are similarly constructed, with the same type of cutting unit (Figure 2.33).

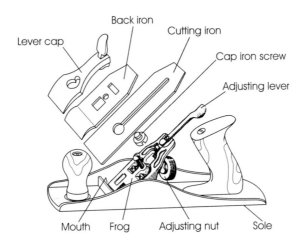

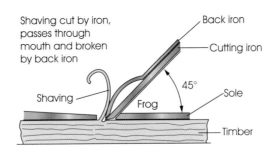

Try plane, rebate plane,
plough plane and chisels (straight)

Jack plane
(slightly round)

Smoothing plane (straight,
corners removed)

Figure 2.33 *Bench plane
exploded view and
cutting action*

Setting

When planing, the thickness of the shaving and the smoothness of the finish is controlled by four main factors (Figure 2.34):

♦ The amount the cutting iron projects below the sole (bottom) of the plane. Hold the plane up and sight along the sole. Turn the adjusting nut, so that the cutting iron projects by about 0.5 mm. Try planing, re-adjust if required to achieve a clean, slightly transparent shaving.

♦ The alignment of the cutting iron should be parallel with the sole. Sight along the sole to check that the cutting iron projects equally right across the mouth (opening in sole). If not it may be corrected by moving the adjusting lever sideways.

♦ The distance the back iron is set from the cutting iron edge. This should be set between 0.5 mm for fine finishing work and up to 2 mm for heavier planing up. To adjust, slacken the cap iron screw (preferably using a wide flat blade screwdriver, not the lever cap!). Set to the required distance and retighten. At this stage it is also worth checking that the back iron sits down closely on the cutting iron without any gaps, otherwise shavings will get jammed between them causing the plane to clog.

◆ The size of the mouth with the cutting iron in position. Between 1.5mm and 3mm is suitable for most timbers. Use a wide mouth for heavy coarse shavings and a narrow mouth for fine work and timber with interlocking grain. The mouth size on bench planes is altered by adjusting the frog. Slacken the frog securing screws, turn the frog adjusting screw in or out to move the frog forwards or backwards, then retighten the frog securing screws.

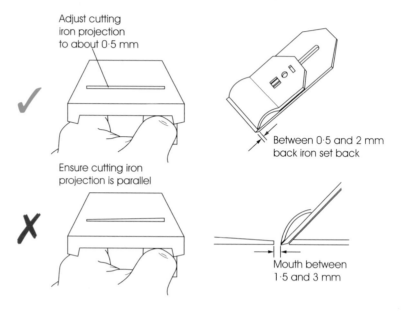

Figure 2.34 *Bench plane settings*

Using bench planes

To accurately plane long boards requires the use of a long jack or try plane (Figure 2.35). A shorter smoothing plane will merely follow the existing bumps and hollows.

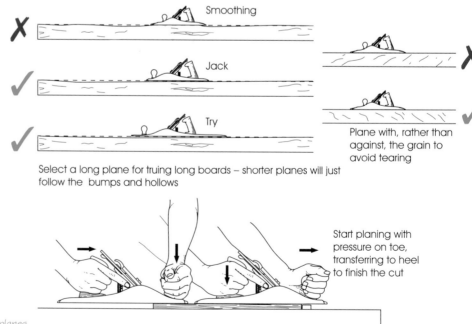

Figure 2.35 *Using bench planes*

Planing requires pressure down on the job as well as moving forward. Begin to plane with pressure on the front or toe of the plane. Move forward over the job, transferring the pressure evenly over the toe and heal.

Finish the cut with pressure on the back or heel of the plane.

Chalk marks can be made across a board before planing, to act as a guide. Chalking three or four lines across a board enables you to see where the plane takes a shaving and what part still needs to be planed.

When cleaning up large areas such as a solid timber table top, use either a jack plane with a slightly curved cutting iron or a smoothing plane where the corners have been removed from the cutting edge. This reduces the chance of creating corner tracks across the work. Use a fine setting and plane diagonally, taking care not to remove too much at the edges (Figure 2.36). Test the surface for flatness by tilting the plane to about 45° using the edge of the sole as a straight-edge. Any light showing through indicates bumps and hollows for further planing.

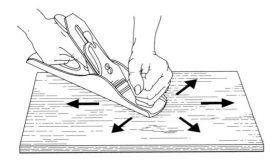

Plane wide tops diagonally

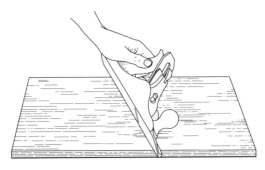

Use edge of sole to check flatness

Figure 2.36 Planing large areas

Planing edges

Edges are normally planed square to a face side (Figure 2.37). Use your plane with the job in the centre of the sole. Steer the plane off to one side or another if you need to correct an angle. A finger curled under the sole at the front acts as a guide and assists steerage.

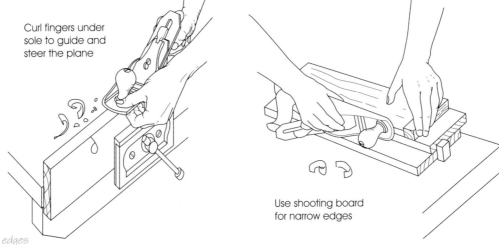

Curl fingers under sole to guide and steer the plane

Use shooting board for narrow edges

Figure 2.37 Planing edges

Handtools Chapter 2

It is often too difficult to balance a bench plane on very thin edges (10 mm or less). A shooting board can be used in these circumstances. The workpiece is supported at right angles to the plane and held against a stop at one end of the shooting board. Lie the plane on its side, using pressure up to the workpiece and down onto the shooting board, sliding the plane along.

Framed joinery

Use a smoothing plane when cleaning up assembled framed joinery (Figure 2.38). A steady turning action towards the joints can help to prevent tearing where the grain direction changes. Tilt the plane across the joints to check for flatness, again any light showing through indicates bumps or hollows, requiring further work.

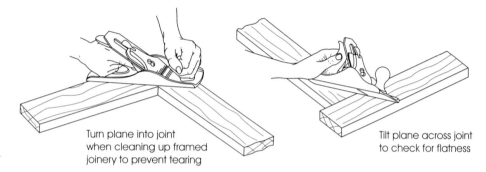

Turn plane into joint when cleaning up framed joinery to prevent tearing

Tilt plane across joint to check for flatness

Figure 2.38 *Cleaning up frame joinery*

Planing end grain

Even if using a sharp, finely set smoothing plane, planing end grain can often result in splintering at the corner (Figure 2.39). This can be avoided using a number of methods:

- By planing from both corners into the centre of the end (not always easy to get a straight edge).
- Trim off the corner as a backing cut and plane towards it.
- Clamp a piece of waste behind the corner and plane towards it. Any splintering will occur in the scrap piece.
- Using a shooting board.

did you know?

Clean up inside edges of all pieces before assembling framed joinery or cabinet carcasses.

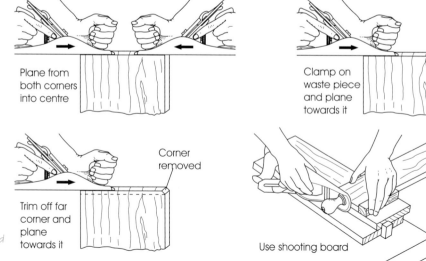

Plane from both corners into centre

Clamp on waste piece and plane towards it

Waste piece

Trim off far corner and plane towards it

Corner removed

Use shooting board

Figure 2.39 *Planing end grain*

Planing sawn timber

Preparing sawn timber by hand can be carried out using the following procedure:

◆ It is normal practice to complete each operation on every piece required for a job before moving on to the next operation.
◆ Select the best faces and direction of working.
◆ Use a jack plane to plane a face side straight and out of twist.
◆ Winding strips can be used to sight along the length as these will accentuate the amount of twist.
◆ Plane diagonally along the board if required to remove any twist.
◆ Mark as face side. (Repeat operation on all other pieces required.)
◆ Use a jack or try plane to plane the face edge straight and square to the face side.
◆ Check periodically with a try square.
◆ Mark as face edge.
◆ Gauge to width from both faces, using the stock of the gauge from the face edge.
◆ Plane down to gauge lines.
◆ Gauge to thickness using the stock of the gauge from the face side.
◆ Plane down to the gauge lines.
◆ Use the blade of a try square or edge of the plane sole to check for flatness across the piece.

Other bench planes

Corrugated sole planes – are versions of jack and try planes (Figure 2.40). They create less friction and are used where resinous timber is regularly encountered. Lubricating the sole of standard bench planes can have a similar effect.

Corrugated sole plane for use with resinous timber

Bench rebate plane has full-width cutting iron for forming and cleaning up large rebates

Pin on a batten to act as a fence when using a bench rebate plane

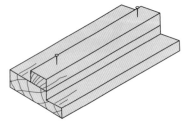

Handtools Chapter 2

Figure 2.40 *Other bench planes*

Bench rebate plane – This is a specialised version of the jack plane with the blade extending the full width of the sole. Also termed a badger or carriage plane, it is mainly used for either forming or cleaning up large rebates. It has no fence or depth stop and is used with a guide batten pinned or cramped to the workpiece.

Specialist planes

With the exception of the compass plane, most specialist planes have a cutting iron only and no back iron. The cutting iron is often positioned with its bevel side up which acts as a back iron to break the wood shaving (Figure 2.41). Various types of specialist planes exist (Figures 2.42 and 2.43).

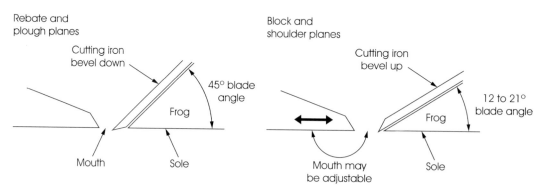

Figure 2.41 *Specialist planes – cutter arrangement*

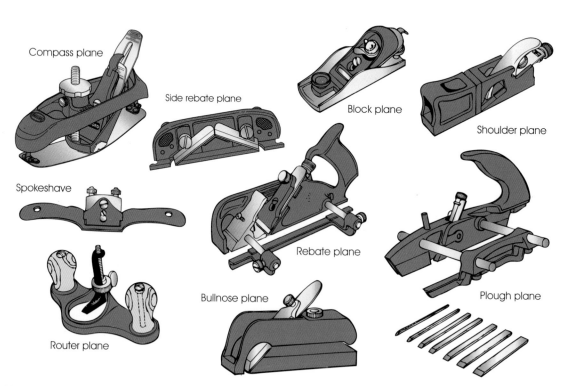

Figure 2.42 *Specialist planes*

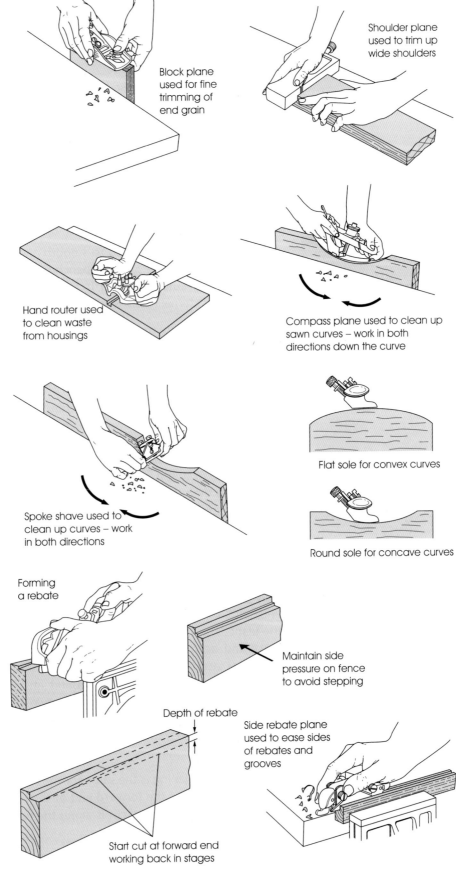

Block plane used for fine trimming of end grain

Shoulder plane used to trim up wide shoulders

Hand router used to clean waste from housings

Compass plane used to clean up sawn curves – work in both directions down the curve

Spoke shave used to clean up curves – work in both directions

Flat sole for convex curves

Round sole for concave curves

Forming a rebate

Maintain side pressure on fence to avoid stepping

Depth of rebate

Side rebate plane used to ease sides of rebates and grooves

Start cut at forward end working back in stages

Figure 2.43 *Use of specialist planes*

Handtools Chapter 2

Block planes – can be used for fine cleaning up and finishing work, both with the grain and for end grain. They are also useful for trimming the edges of manufactured boards and plastic laminates. Versions are available with low angled blades and adjustable mouths to fine tune their cutting efficiency.

Shoulder planes – are used for cleaning up and truing rebates and the end grain shoulders of tenon joints, etc.

Bull-nose planes – are shortened versions of a shoulder plane. Their cutting iron is mounted towards the front, so that they can work close into corners, such as stopped rebates and chamfers.

Router planes – often termed grannies tooth – are used to clean out the waste wood from housings and grooves. Various blade widths are available to suit the work in hand and the depth of cut is adjustable.

Compass planes – are used to smooth and clean up sawn curves. Their cutting assembly is the same as bench planes, but the sole is made from a flexible steel strip, enabling it to be adjusted to fit against internal (concave) or external (convex) curved shapes.

Spoke-shaves – are used for the final working and cleaning up of curved edges. The flat-soled version can be used for planing narrow edges and convex curves, concave curves being worked with a round-soled spoke-shave. In use, frequent changes of direction may be required.

Rebate planes – are used to cut rebates. The size of rebate is controlled by the use of an adjustable side fence for width and a depth stop. In use, the fence must be firmly held against the face or edge of the work to prevent 'stepping'. The cutting iron has an additional mounting position at the front, so that it can work close into the corners of stopped rebates. A spur near the sole can be adjusted to cut the grain when planing rebates across the grain.

Side rebate planes – are used to widen and clean up the sides of grooves, rebates and housings. They have two opposing blades, enabling both sides of a groove to be eased in the same direction with the grain.

Plough planes – are used to cut plough grooves both along the grain and, with a spur, across the grain. A range of blade widths is supplied with the plane. They may also be used for rebates. Before each use, set up the fence and depth stop, ensuring the fence is parallel to the plane body. In use, as with rebate planes, the fence must be firmly held against the workpiece to prevent stepping. Rebates and plough grooves wider than the blades supplied may be formed in two or more stages. Multi-planes similar in appearance to plough planes are available; these are supplied with an extended range of cutters, allowing rebating, ploughing and moulding operations.

The method of planing is the same for the rebate, plough and multi-plane. The cut is started at the forward end of the piece and the plane gradually worked back until the final cut is along the full length.

did you know?

Work with the grain when planning curves, for a clean finish.

did you know?

To avoid tearing always work with the grain.

Sharpening planes and chisels

Plane cutting irons (blades) and chisels are ground to a bevel at the 'sharp' end. An angle of 25° is the most suitable for general woodwork. A slacker angle results in a thin edge, which is easily damaged; any steeper and cutting will be difficult.

A secondary bevel of 30° is honed on the very edge of the grinding bevel to strengthen the bevel and provide a fine cutting edge (Figure 2.44).

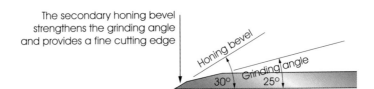

The secondary honing bevel strengthens the grinding angle and provides a fine cutting edge

Honing bevel

Grinding angle

30° Grinding 25°

Figure 2.44 *Sharpening and shaping cutting edges*

Grinding

The sharpening of cutting irons and chisels is therefore a two-stage process. Grinding is carried out on one of the following:

◆ Using a sandstone is a wet grinding process because water is used to keep the cutting iron cool and to prevent the stone becoming clogged.

◆ Using a high speed carborundum grinding wheel is a dry grinding process, although to prevent the cutting iron overheating, it can be periodically cooled in water. If the tool is allowed to overheat it will lose its hardness.

◆ A coarse sharpening stone is used when a grinding wheel is not available.

Honing

This is carried out on a fine sharpening stone. The fine side is for honing, the coarse side is only used when a grindstone is not available.

These are available in many materials.

Oilstones – such as natural Arkansas, synthetic silicon-carbide. Coarse medium and fine grades are available. In addition, combination stones are available, which have two different grade stones glued back to back. Use the finer side for honing and the coarser side to occasionally replace the grinding bevel. As their name suggests, oil is normally used to lubricate the stones and prevent them clogging with metal particles.

Japanese waterstones – whether natural or synthetic, these are quick cutting and available in finer grades than oilstones, but are more expensive. They are lubricated with water; it helps to build up 'a slurry' on the wet surface prior to honing by rubbing the surface with a chalk-like Nagura stone.

Diamond stones – have a grid-like pattern of diamond particles bonded in a plastic base. They are available in various grades, but are more expensive. Diamond stones can be used to flatten out worn oil and water stones.

Slipstones – both oil- and waterstones are available in a teardrop section for honing inside ground gouges.

Cutting edge shape

The shape of a blade's cutting edge will depend on its use. Most will be square. The corners of a smoothing plane blade are removed to prevent them digging in and ridging the work surface. Jack plane irons are slightly rounded to prevent digging in and aid the quick, easy removal of shavings.

Sharpening procedure

did you know?

When working difficult grains some woodworkers further hone a cutting iron by working the bevel on a thick piece of leather.

◆ Grind iron when required on a grinding wheel with the tool rest set at an angle of 25° (Figure 2.45). Test for squareness.
◆ Hone iron on the fine side of a lubricated sharpening stone. Place the grinding bevel flat on the stone. Lift slightly and use firm to and fro strokes to form the honing bevel. Wide plane blades may require angling across the stone, to avoid forming a hollow. Continue honing until a bevel about 1 mm wide is formed.
◆ The honing process forms a burr on the underside of the iron. Check for it by running your thumb over the back of the blade. Use light strokes to remove the burr with the back of the iron flat on the stone.
◆ Finally, remove the thin wire edge left after honing by drawing the cutting edge across a piece of wood.

safety tip

Always use eye protection when using a grinding wheel.

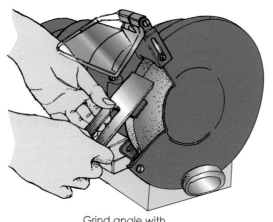
Grind angle with
tool rest set at 25°

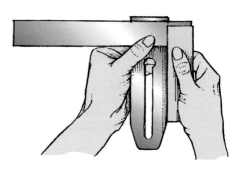
Use try square to
test for squareness

Use firm to-and-fro strokes or a figure-
of-eight pattern to form honing bevel

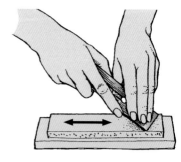
Wide blades may require
angling across the stone

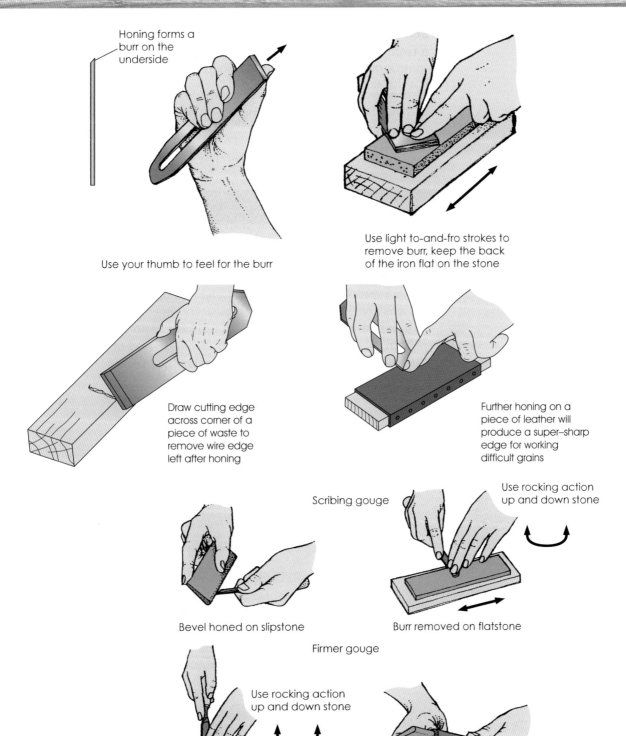

Honing forms a burr on the underside

Use your thumb to feel for the burr

Use light to-and-fro strokes to remove burr, keep the back of the iron flat on the stone

Draw cutting edge across corner of a piece of waste to remove wire edge left after honing

Further honing on a piece of leather will produce a super–sharp edge for working difficult grains

Scribing gouge

Use rocking action up and down stone

Bevel honed on slipstone

Burr removed on flatstone

Firmer gouge

Use rocking action up and down stone

Bevel honed on flatstone

Burr removed on slipstone

Figure 2.45 *Sharpening procedure*

Sharpening gouges

A similar procedure is used to sharpen gouges, except as they are curved-blade chisels, shaped grindstones and slip-stones are required.

Scribing gouges – are ground on a shaped grindstone and honed using a teardrop-shaped slipstone. The burr is removed on a standard flatstone, using a rocking action as it is moved to and fro along the stone.

Firmer gouges – can be ground and honed on standard grinding wheels and flatstones, using a rocking action. A teardrop slipstone is used to remove the burr.

Chisels and gouges

Firmer chisels – are general-purpose chisels, which can be used for all types of woodwork (Figure 2.46). They have a rectangular section blade, which is strong enough to be driven through wood with preferably the aid of a wooden mallet or the flat (side) face of the hammer. Firmer blades are available in widths rising in regular increments from 3 mm to 30 mm; 45 mm and 50 mm blades are also available in some ranges. Registered pattern chisels have a steel band (ferrule) at the end of the handle to stop the wooden handle splitting under constant hammering.

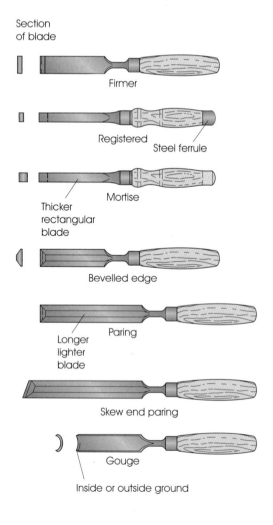

Figure 2.46 *Chisels and gouges*

Mortise chisels – are much stronger than firmer chisels with a thicker section rectangular blade to withstand the heavy hammering and levering involved in chopping mortises. Standard widths are 6, 9 and 12 mm. Other sizes up to 25 mm wide are available in some ranges.

Bevelled edge chisels – are a lighter form of firmer chisel with bevels on the front edges of the blade. The thinner edges of this chisel enable it to be used for chiselling corners, which are less than 90°, such as dovetails, etc. They are available in widths ranging from 3 mm to 50 mm.

Paring chisels – are similar to bevelled-edge chisels, but are much longer and lighter. They must be used for handwork only and never hit with a mallet. Their extra length makes them easier to control when paring either vertically or horizontally. They are available in various widths up to 50 mm. Skew end paring chisels are also available for paring into awkward corners.

Gouges – are in fact curved chisels, and mainly used for shaping, scribing and carving. Two types are available:

◆ a scribing gouge is ground and honed on its inside curvature for paring and scribing concave surfaces;
◆ a firmer gouge is ground and honed on its outside curvature for hollowing and carving.

Using chisels

did you know?

Always keep both hands behind the cutting edge when paring with a chisel.

The most important point when using a chisel is to ensure that it is sharp. For much of the time you will be using hand pressure – when extra pressure is required re-hone it to restore the edge.

Paring

This is the cutting of thin slices of wood, either across the grain's length or across the end grain (Figure 2.47).

Vertical end grain paring –
◆ Place the workpiece flat on the bench. A bench hook or cutting board should be used to protect the bench surface.
◆ Grip the chisel handle with your thumb over the end.
◆ Control the blade with the thumb and forefinger of your other hand.
◆ With the chisel vertical and close to your body, use your shoulder power to apply a steady downwards force.
◆ Work from the corner in towards the centre, paring off a little wood at a time.

Horizontal across the grain paring –
◆ Secure the workpiece to the bench top.
◆ Use a tenon saw to cut the sides of the housing. Wide housings may require additional saw cuts in the waste.
◆ Grip the chisel handle with your index finger extended towards the blade.
◆ Use the thumb and forefinger of your freehand to grip and guide the blade.

safety tip

Never use your palm to strike the end of the handle as it will cause personal injury in the long term.

◆ Take up a position in front of the workpiece with your forearm and the chisel parallel with the floor. Use your body weight to push the chisel forward.

◆ Work from one side towards the centre, taking a little wood off at a time.

◆ Turn the workpiece around and complete from the other side. Where extra force is required a mallet may be used.

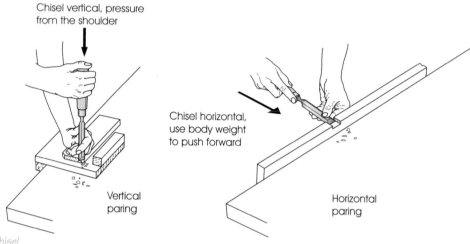

Chisel vertical, pressure from the shoulder

Chisel horizontal, use body weight to push forward

Vertical paring

Horizontal paring

Figure 2.47 *Paring with a chisel*

Chopping with a chisel and mallet –

◆ Secure the workpiece to the bench top.

◆ Grip the chisel handle through a clenched fist (Figure 2.48).

◆ Hold the chisel vertical and strike the end of the handle squarely with the face of the mallet for maximum force.

◆ For more delicate chopping hold the mallet shaft just below the head and lightly tap the chisel with the side of the mallet.

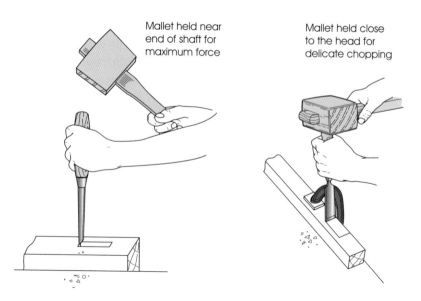

Mallet held near end of shaft for maximum force

Mallet held close to the head for delicate chopping

Figure 2.48 *Chopping with a chisel and mallet*

Paring and chopping with a gouge – requires the same techniques used for standard chiselling. Use a scribing gouge for paring internal concave, curved shoulders. The firmer gouge is used for scooping out hollows in the surface of the workpiece.

Drills and braces

With the increasing availability of portable power tools and especially battery-powered drills, the use of hand drills and braces is today far less commonplace. However, they may be still a useful addition to the toolkit as they are simple, safe and the only power source required is yourself.

Wheel brace and twist drill

Also termed a hand drill, it is used both on site and in the workshop (Figure 2.49). Its main use is for boring pilot holes for screws, etc. Twist drills are used in this type of brace. They are available in a range of sizes from 1 mm to 6 mm in diameter. A countersink bit is also available for the wheel brace.

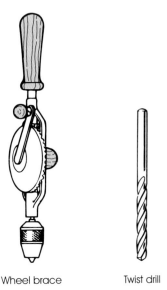

Wheel brace   Twist drill

Figure 2.49 *Wheel brace and twist drill*

Ratchet brace and bits

These are used mainly for site work (Figure 2.50). The handle of the brace has a sweep of 125 mm. Sometimes it is necessary to bore holes in or near corners. In order to do this the ratchet on the brace is put into action.

There is a wide range of bits available:

◆ Jennings bit (6 mm to 38 mm in diameter) – used for boring both across the grain and into end grain.
◆ Centre bit (6 mm to 32 mm in diameter) – used for boring shallow holes.
◆ Irwin bit (6 mm to 38 mm in diameter) – has the same range of uses as the Jennings bit.

◆ Forstner bit (10 mm to 50 mm in diameter) – used for cutting blind or flat-bottomed holes.
◆ Expanding bits (expands 13 mm to 75 mm) – used for boring large holes.
◆ Countersink bits – used to prepare holes in order to receive countersunk screws.
◆ Screwdriver bits – can be used as an alternative to the hand screwdriver, giving more leverage and a quicker action.

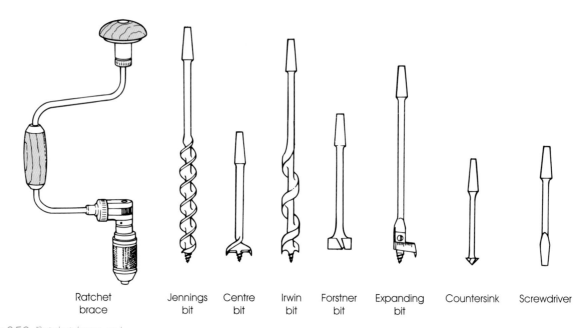

Ratchet brace   Jennings bit   Centre bit   Irwin bit   Forstner bit   Expanding bit   Countersink   Screwdriver

Figure 2.50 *Ratchet brace and bits*

Use of drills

The most difficult part of hand drilling or boring is to start and keep the drill or bit going in the right direction. Good all-round vision is essential. An assistant can be useful until the skill is mastered, as they can tell you whether the drill is leaning out of line horizontally, vertically or parallel with the edge (Figure 2.51).

◆ Small holes up to say 6 mm can be drilled using a twist drill in a hand or power drill.
◆ Larger holes up to about 50 mm can be bored either using a handbrace and auger bit or a spade bit in a hand or power drill.

In order to prevent splitting out the workface when boring through holes either:

◆ bore from one face until the point of the bit appears, then complete the hole from the other side placing the point of the bit into the small hole and complete the work; or
◆ temporarily cramp a waste piece of wood on the back of the workpiece and bore right through. Any splitting will then be in the waste and not the workpiece.

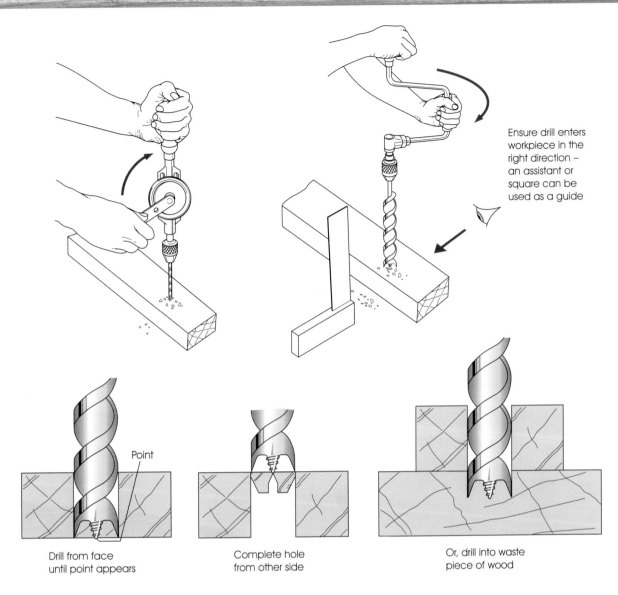

Ensure drill enters workpiece in the right direction – an assistant or square can be used as a guide

Point

Drill from face until point appears

Complete hole from other side

Or, drill into waste piece of wood

Figure 2.51 *Using a drill*

Fixing tools

Hammers

Hammers are available in three main types for use by the woodworker (Figure 2.52):

◆ Warrington or cross-pane hammer – used mainly by the joiner for shop work.
◆ Claw hammer – a heavier hammer than the Warrington; used mainly by the carpenter for site work.
◆ Pin hammer – a lighter version of the Warrington; used both on site and in the workshop for driving small pins.

did you know?

Gripping a hammer or mallet shaft at the end furthest from the head requires less effort and gives better control.

Hammer handles or shafts – traditionally available in hickory or ash. Today steel and glass-fibre are more common. Hickory is the best wooden handle, but the steel or glass-fibre shafted hammer is often preferred for its balance and increased strength.

Mallets – are mainly a joiner's tool, used for driving chisels when cutting joints and for framing up. A suitable material for the head of a mallet is beech and the shaft is normally made from beech or ash. Rubber-headed mallets are useful for frame and cabinet assembly, as they are less likely to damage the work.

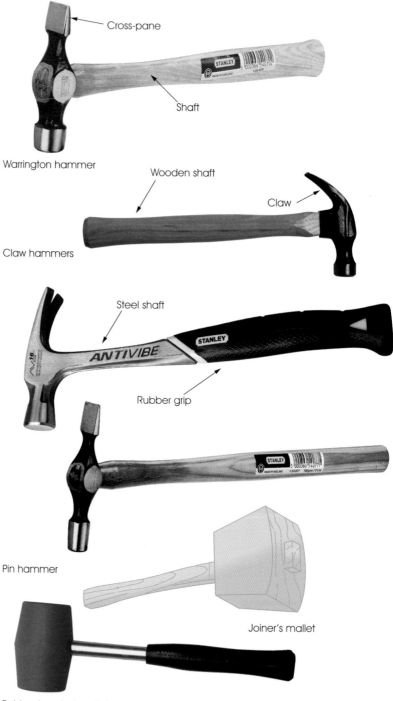

Cross-pane

Shaft

Warrington hammer

Wooden shaft

Claw

Claw hammers

Steel shaft

Rubber grip

Pin hammer

Joiner's mallet

Rubber-headed mallet

Figure 2.52 *Hammers and mallets*

Use of hammers and mallets

◆ Grip the hammers and mallets at the end of the shaft, not near the head. Less effort and better control is achieved in this way.
◆ Use the cross-pane to start small pins.
◆ Use a nail punch to drive nail heads below the surface to avoid bruising the work.
◆ Keep the hammerhead clean to prevent slipping and bending nails.
◆ Small bruises in the workpiece caused by a slipping hammer may be raised by immediately locally soaking the bruise with water to swell the wood; application of a hot iron will speed the process. Once the fibres are raised to the surface, allow them to dry and then sand off smooth.
◆ Place a scrap block of wood under a claw hammer or pincers when withdrawing nails. This protects the workpiece and increases leverage.

Screwdrivers

There are four main types of screwdriver used by the woodworker (Figure 2.53).

Cabinet screwdriver – traditionally the cabinet pattern screwdriver has a rounded steel blade and wooden handle. The London pattern has a flat blade and flat faces on the handle. Both are available in sizes (length of the blade) ranging from 50 mm to 300 mm and various blade widths for slot-head screws.

Ratchet screwdriver – is a handy tool as it allows screws to be turned in or removed without releasing the handle. It is available in a similar range of sizes to those of the cabinet screwdriver. Both slot-head and cross-head types are available.

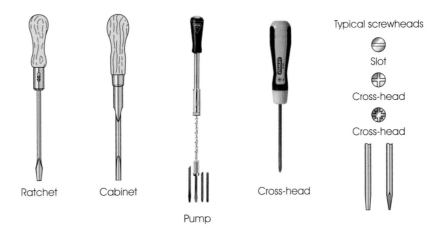

Figure 2.53 *Screwdrivers*

Ratchet   Cabinet   Pump   Cross-head

Typical screwheads
Slot
Cross-head
Cross-head

Pump screwdriver – popular with both carpenters and joiners as it allows quick and easy insertion and removal of screws. A range of interchangeable screwdriver bits, drill bits and countersinks are available for this type of screwdriver. In use, take a grip at the chuck end to steady the screwdriver bit and to prevent it from jumping out of the screw head and damaging the workpiece.

Handtools

Chapter 2

Cross-head screwdriver – are available in a range of blade lengths, tip patterns and sizes to suit the variety of cross-head screws manufactured.

Finishing tools and abrasives

Scrapers

Scrapers remove paper-thin shavings, leaving a smooth finish, even on irregular grain. Cabinet scrapers are a section of hardened steel sheet, either rectangular for flat surfaces or shaped for finishing mouldings and other shaped work (Figure 2.54). A burr is turned on the edge of the scraper to form the cutting edge.

◆ Hold the scraper in both hands using your thumbs to flex it into a curve.
◆ Tilt the scraper away from you using a push forward action along the work.
◆ Vary the tilt and the flex so that you are removing the required amount.
◆ For wide surfaces use diagonal strokes in both directions before finishing parallel with the grain.

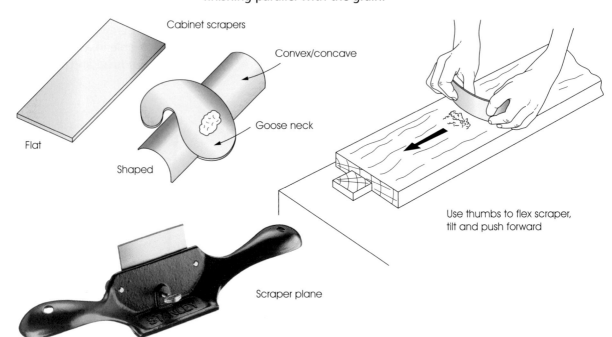

Cabinet scrapers

Convex/concave

Goose neck

Flat

Shaped

Use thumbs to flex scraper, tilt and push forward

Scraper plane

Figure 2.54 *Scrapers*

Scraper plane

Flexing a cabinet scraper can get hard on the thumbs after a period. The blade of a scraper plane is held in the plane body at the required cutting angle and is flexed into a curve by an adjustment screw. The plane is simply pushed forward. Adjust the curve until the required shaving is being made. Slack curves will only produce dust. Greater curves produce thicker shavings.

Sharpening a scraper

◆ Secure the scraper in a vice and 'draw file' its long edges true and square (Figure 2.55), using your fingertips to prevent the file rocking as you pull it along the scraper.

◆ Hone the filed edge by drawing an oiled slipstone over it. Use the stone on each side to remove the burr.

◆ To raise a burr, lay the scraper on the bench top, with the long edge just overhanging. Stroke the face at a slight angle along the entire edge four or five times using a burnisher (like a toothless file) or the curved back of a gouge blade. Repeat from the other face.

◆ To turn a burr, hold the scraper on end, with the burnisher or gouge held at a slight angle, and make two or three firm strokes from both sides.

◆ When the scraper starts to dull, simply raise and turn a new burr. Draw filing and honing is only required periodically when the edge has been damaged.

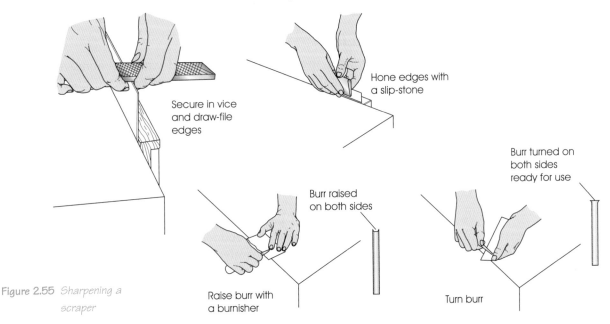

Secure in vice and draw-file edges

Hone edges with a slip-stone

Burr raised on both sides

Raise burr with a burnisher

Burr turned on both sides ready for use

Turn burr

Figure 2.55 *Sharpening a scraper*

Sanding

This is the process of finishing the surface of wood with an abrasive paper or cloth. The method used will depend on the surface.

Hardwoods – are normally clear finished. Planed surfaces can be scraped to remove minor blemishes and torn grain before finally rubbing down, parallel with the grain, with an abrasive paper.

Softwoods – are normally painted. Minor blemishes will not show through. Machine marks should be removed with a smoothing plane, followed by rubbing down diagonally. The small scratches left as a result help to form a key for the priming paint.

Abrasive papers

A variety of materials are bonded to backing sheets. Traditionally the most popular for hand sanding wood was glass and garnet (harder than glass).

did you know?

Open-coated abrasive papers are less likely to clog up with wood dust than are closed-coated abrasive papers.

Handtools **Chapter 2**

Increasingly, aluminium oxide (harder than garnet) and silicon-carbide (wet and dry paper) are being used.

Most are available in standard sheet sizes of 280 mm × 230 mm. Smaller cut sheet sizes are provided to suit standard power sander bases and in rolls of various widths.

Abrasive papers are graded according to the size of particle (grit size), typically termed very coarse, coarse, medium, fine or very fine. They are also graded by number, typically from 50 to 400 or 1 to 9/0, the higher the number the finer the grit.

In addition to grades, abrasive papers are also termed as either open-coated or closed-coated depending on the spacing of the particles. The particles of close-coated abrasives are bonded closely together for fast, fine finishing, whereas in open-coated the abrasive particles are spaced further apart. These are coarser, clog up less readily and are thus better for use when finishing resinous timbers.

Hand sanding

- Flat surfaces should always be sanded with the abrasive wrapped around a sanding block to ensure a uniform surface (Figure 2.56).
- Mouldings may be sanded with the abrasive wrapped around a shaped block.
- Curved surfaces and rounded edges may be sanded without a block. Use your palm or finger tips to apply pressure to the abrasive.
- Arrises (the sharp external edge where two surfaces meet) are removed (de-arrised) using the abrasive wrapped around a block. The purpose of de-arrising is to soften the sharp edge and provide a better surface for the subsequent paint or polished finish. However, take care not to overdo it by completely rounding over a sharp corner. All that is required is a very fine chamfer. Run your thumb over a number of arrises and you will soon get the 'feel' for the correct profile.

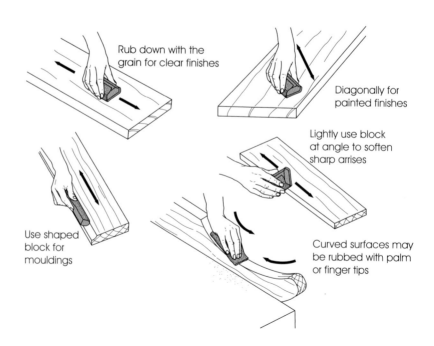

Rub down with the grain for clear finishes

Diagonally for painted finishes

Lightly use block at angle to soften sharp arrises

Use shaped block for mouldings

Curved surfaces may be rubbed with palm or finger tips

Figure 2.56 Hand sanding

Miscellaneous tools

Pincers – are used both on site and in the shop for removing small nails and pins (Figure 2.57). To avoid bruising the wood, joiners place a scraper or a thin block of timber under the levering jaw.

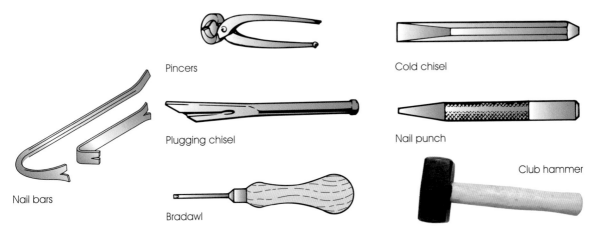

Pincers

Cold chisel

Plugging chisel

Nail punch

Club hammer

Nail bars

Bradawl

Figure 2.57 *Miscellaneous tools*

Cold chisels – are steel chisels used for cutting holes or openings in brickwork, etc.

Plugging chisels –are also steel chisels but exclusively used for raking out the mortar joints in brickwork to receive wooden plugs and joist hangers, etc.

Nail punches – are used to set nails below the surface of the timber. They are available in a wide range of sizes to suit most nails.

Bradawls – are used to form starting holes for nails and small screws. Pilot holes for larger screws should always be bored with a twist drill.

Club hammers – are also known as lump hammers. They are mainly used by bricklayers but are useful to the woodworker for heavy work when using plugging chisels, etc.

Nail bars – also known as wrecking bars, tommy bars or crow bars. Used for levering, prizing items apart and withdrawing nails.

Rasps – are rough woodworkers' files, used for roughing out shaped work (Figure 2.58).

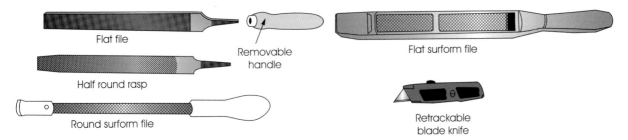

Flat file

Removable handle

Flat surform file

Half round rasp

Round surform file

Retrackable blade knife

Figure 2.58 *Files, rasps and knives*

Handtools **Chapter 2**

Metal-worker's files – are used for the maintenance of tools and adjusting metal fittings, etc. They are also useful for the final finishing of edges to plastic laminate. In use, files and rasps can become clogged and no longer cut: they should be cleaned using a file card or coarse wire brush.

Surform files – are a development of the rasp. They have thin perforated metal blades that allow wood shavings to pass through without clogging. There is a wide range available, suitable for planing and roughing-out shaped work.

Replaceable blade knives – accept various blades enabling them to be used for cutting plasterboard, plastic laminate, building paper, roofing felt and insulation board, etc. They can also be fitted with a pad-saw blade for cutting timber, or a hacksaw blade for cutting metal.

Workshop equipment

A large number of items are required in order to produce a finished piece of work. The main equipment include (Figure 2.59):

- bench
- stools
- cramps
- bench equipment
- templates.

Bench

The ideal woodworkers' bench should be 800 mm high, 3 to 4 metres in length and have a well or recess running down its centre so that tools may be placed in it without obstructing the bench surface. The bench should also have a bench stop, bench peg and a steel instantaneous-grip vice which opens at least 300 mm.

Saw stool

Most woodworkers require one or two sawing stools for their own use. These should be sturdily constructed with 50 mm × 50 mm legs, housed into a 75 mm × 100 mm top.

Cramps

Many types of cramp or clamp are used by the woodworker:

Sash cramp – available in sizes from 500 mm to 2.0 m. This cramp is used to pull up joints, etc. when assembling and gluing up.

Gee cramp – available in sizes from 100 mm to 300 mm. This cramp is used for general holding and cramping jobs before and after assembly. Edging and deepthroat versions are also available.

Holdfast – used for cramping jobs to the bench for sawing, cleaning up and finishing. The leg of the holdfast locates in a hole in the bench top.

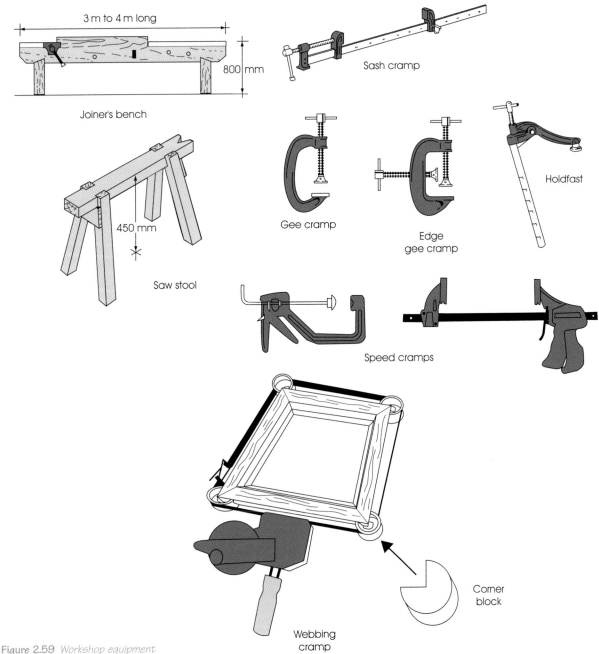

Figure 2.59 *Workshop equipment*

Speed cramps – available in a variety of types. All enable rapid single-handed cramping whilst the workpiece is held in the other hand. Mainly used in place of gee cramps.

Long bar ratchet cramps – may be used in place of sash cramps when assembling small frames. However, it is almost impossible to apply the same pressure.

Webbing cramps – also termed picture frame cramps, as they are useful in the assembly of small mitred frames, and can also be used to cramp irregular shapes. They consist of a length of webbing, which is pulled taut around the workpiece by a ratchet mechanism. Corner blocks may be used to apply extra pressure at that point.

Handtools Chapter 2

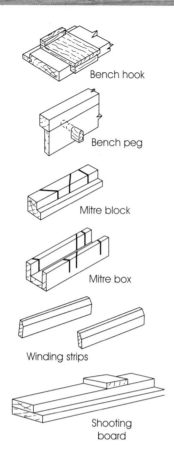

Bench hook

Bench peg

Mitre block

Mitre box

Winding strips

Shooting board

Figure 2.60 Bench equipment

Bench equipment

Most woodworkers will make various items of wooden equipment for themselves. These may include the following items (Figure 2.60):

Bench hook – used to steady small pieces of timber when cross-cutting, e.g. squaring ends, cutting to length and sawing shoulders.

Bench pegs – can be inserted in various positions along the side of the bench to support longboards, etc.

Mitre block – used when cutting mitres for picture frames and other small mouldings.

Mitre box – used when cutting mitres on larger mouldings such as architraves and skirtings.

Winding strips – used to check timber and framed joinery items for wind or twist.

Shooting board – used as an aid when planing end grain and thin boards. An angled stop can be use when trimming mitres.

Templates

Woodworkers often make up jigs and templates in order to make the job easier (Figure 2.61). Some of these are described below.

Mitre template

Dovetail template

Box square

Figure 2.61 Templates

◆ Dovetail template – used to mark out dovetails.
◆ Mitre template – used to guide the chisel when mitring or scribing joints.
◆ Box square – used to square lines around moulded sections.

 Site equipment

Straight-edge – A straight length of timber or an off-cut from the factory machined edge of sheet material. It is used on its own to test the alignment of spaced members or in conjunction with a spirit level for plumbing and levelling (Figure 2.62).

Builder's square – A purpose-made large square used for setting out right angles such as partitions and walls. It can be cut from the corner of a sheet or jointed together. Use the 3, 4, 5 rule to set it out, typically make the sides 900, 1200 and 1500 mm.

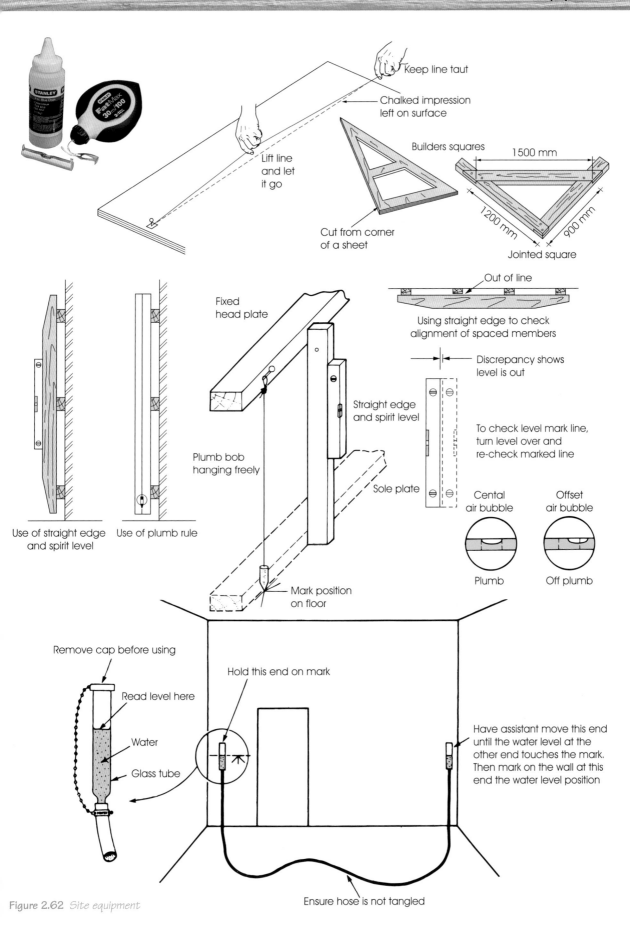

Keep line taut

Chalked impression left on surface

Lift line and let it go

Builders squares

1500 mm

Cut from corner of a sheet

1200 mm

900 mm

Jointed square

Fixed head plate

Out of line

Using straight edge to check alignment of spaced members

Discrepancy shows level is out

Straight edge and spirit level

To check level mark line, turn level over and re-check marked line

Plumb bob hanging freely

Sole plate

Central air bubble

Offset air bubble

Use of straight edge and spirit level

Use of plumb rule

Plumb

Off plumb

Mark position on floor

Remove cap before using

Read level here

Water

Glass tube

Hold this end on mark

Have assistant move this end until the water level at the other end touches the mark. Then mark on the wall at this end the water level position

Ensure hose is not tangled

Figure 2.62 *Site equipment*

Handtools

Chapter 2

Spirit levels – Traditionally they had a hardwood body, but are now almost exclusively aluminium. Inset in the body are a number of curved tubes, partially filled with spirit and containing a bubble of air. The bubble position indicates whether a surface is truly plumb (vertical) or level (horizontal). Before use, check the level for accuracy. Use the level to mark a pencil line on a wall, either vertical or horizontal. Turn the level over and re-check the line. Any discrepancy and the level is 'out'. The tubes may be re-set in some levels, otherwise it is time to get a new one. Tripod-mounted laser versions of the sprit level are now fairly inexpensive; these can be used to project level datum points around a room.

Water levels – Consist of a length of hose with a glass or transparent plastic tube at each end. The water surfaces in the two tubes give two level points. This uses the fact that water will always find its own level. It is used mainly to transfer levels over a distance or around corners, etc.

Plumb bob – This is a metal weight attached to a length of cord. When freely suspended, it produces a true vertical plumb line. It is used to indicate drop positions or as a margin line from which other positions can be measured.

Plumb rule – This is made using a straight-edge and plumb bob. It allows the rule edge to touch the item being plumbed or tested. A hole is cut in one end of the edge to accommodate the bob and a central gauge line is used as the plumb indicator.

Chalk line – A line coated with chalk used to mark straight lines on a surface. The line is stretched over a surface and snapped in the centre, leaving a chalked impression. Purpose-made chalk lines are available, consisting of a container filled with coloured chalk into which the line is wound. In addition these can often double up as plumb bob, enabling vertical lines to be chalked.

activity

Make a list of hand tools and equipment required to hang an inward opening external front entrance door to a client's house, including the fitting of related items of door furniture. Refer to Chapter 6.

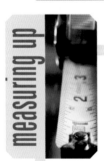

measuring up

1. State why 'open-coated' abrasive papers are best when finishing resinous timbers.

2. Name the type of saw tooth best suited for cutting along the grain.

3. Explain in sequence, the procedure for sharpening a chisel.

4. Describe what is meant by 'turning the burr' and name the tool it applies to.

5. State what has gone wrong during rebating, as shown in the illustration.

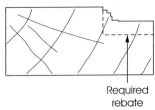

Required
rebate

6. State how the accuracy of a spirit level may be checked.

7. Name THREE different types of plane and state the use for each.

8. State why the corners of a smoothing plane blade may be removed after sharpening.

9. List in the sequence they are carried out the operations involved in sharpening a saw.

10. Produce sketches to show the difference between a Warrington and a claw hammer. State typical uses for each.

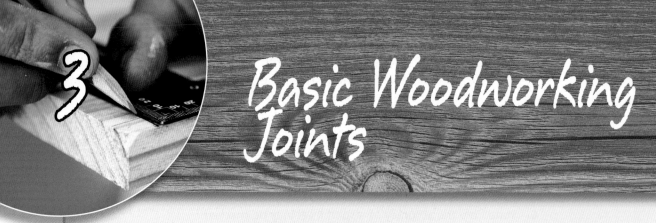

Basic Woodworking Joints

This chapter is intended to provide the reader with an overview of the types of joint, their proportions and assembly methods, which are used by the woodworker. Although its content is not assessed directly, knowledge of its contents is assumed and assessed in Wood Occupations units VR 05, VR 06, VR 07, and VR 08 at Level 1. Knowledge is also assessed in both Site Carpentry and Bench Joinery at Level 2.

In this chapter you will cover the following range of topics:

- Butt joints and mitres
- Lap joints
- Halving joints
- Housing joints
- Edge joints
- Notched and cogged joints
- Dovetail joints
- Mortise and tenon joints
- Dowel joints
- Assembling frames
- Board fittings, edging and laminating

What's required in VR 05, VR 06, VR 07 & VR 08?

To successfully complete these units you will be required to demonstrate your skill and knowledge of the following processes:

- Marking out and forming basic woodworking joints using a range of woodworking hand tools.

You will be required practically to:

- Mark out and form halving, housing, notched, mortised, bridle, edge and angle joints.
- Work with a range of softwoods, hardwoods, wood-based sheet materials, adhesives and associated fixings.
- Use a range of woodworking handtools for sawing, planing, chiselling, shaping, drilling, screwing, hammering, marking and testing.
- Use a range of workshop and cramping equipment.

All types of woodworker make joints. The carpenter makes joints that are normally load bearing. The joiner uses mainly framing joints for doors, windows and decorative trims. The cabinet/furniture maker also uses framing joints, both to create flat frames and also to build up three-dimensional carcasses. In addition both the joiner and the cabinet/furniture maker may be involved with a range of jointing methods for manufactured boards.

Joint making is regarded as a measure of a woodworker's skill, since it requires the mastering of a variety of very accurate marking and cutting techniques. In training, joint making instils a 'feel' for both materials and the use of hand tools, which will not be lost in later years, even when progressing to machines and powered hand tools.

The most common hand cut joints are described here, followed by the processes required to form them.

Butt joints and mitres

Butt joints are the simplest form of joint, where one piece of timber meets another. It is not a strong joint on its own as end grain does not take glue well and there are no interlocking parts. However, they are often reinforced in some way for increased strength.

Butt joints in length – are mainly used when joining structural timber and covering material (Figure 3.1). They must always be made over a support, for example, wall plate or joist.

Butt joints for boxes and frames – are simple angle joints (Figure 3.2) made with the end grain of one glued to the face or edge of the other.

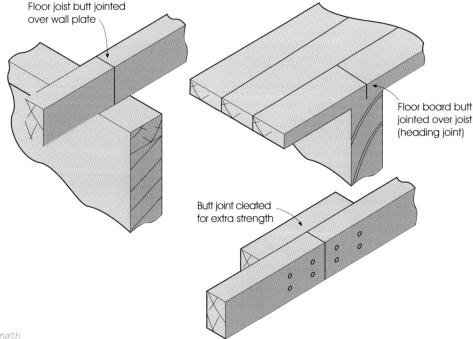

Floor joist butt jointed over wall plate

Floor board butt jointed over joist (heading joint)

Butt joint cleated for extra strength

Figure 3.1 *Butt joints in length*

Basic Woodworking Joints **Chapter 3**

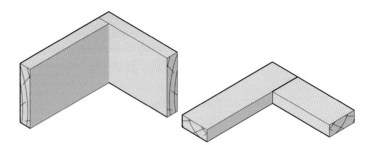

Figure 3.2 *But joints in boxes and frames*

Forming butt joints

◆ Mark the length of the parts and with a try square and marking knife; square a shoulder line across the face and edge (Figure 3.3).
◆ Use a bench hook to hold the piece; cut square on the shoulder line using a tenon saw.
◆ Trim the end grain to the shoulder line using a plane and shooting board to provide a smooth surface for gluing.
◆ Apply glue to both parts, rub together to expel excess glue and cramp together.
◆ Use a damp cloth to wipe off the excess glue.

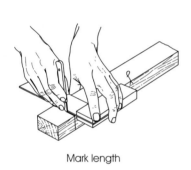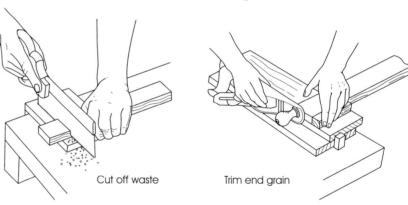

Mark length   Cut off waste   Trim end grain

Figure 3.3 *Forming a butt joint*

Reinforcing butt joints

As end grain does not glue well, the joint is often reinforced with nails or glue blocks (Figure 3.4).

Butt joints can also be screwed or dowelled for additional strength.

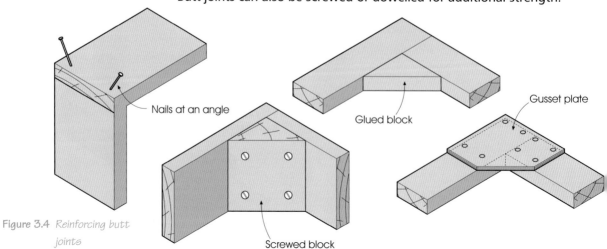

Nails at an angle   Glued block   Gusset plate

Screwed block

Figure 3.4 *Reinforcing butt joints*

Mitred butt joint

The use of mitres increases the gluing area (compared with a squared-end butt) avoids end grain being visible and also allows moulded sections to be joined. In addition to gluing, the mitre is normally reinforced by nailing (Figure 3.5) or by using splines or tongues (Figures 3.7 and 3.8).

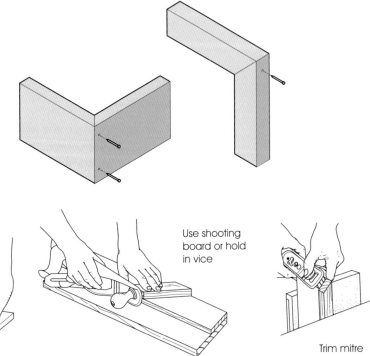

Figure 3.5 *Mitre joints*

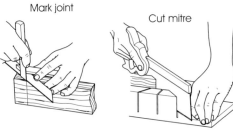

Mark joint

Cut mitre

Use shooting board or hold in vice

Trim mitre

Figure 3.6 *Forming a mitred butt joint*

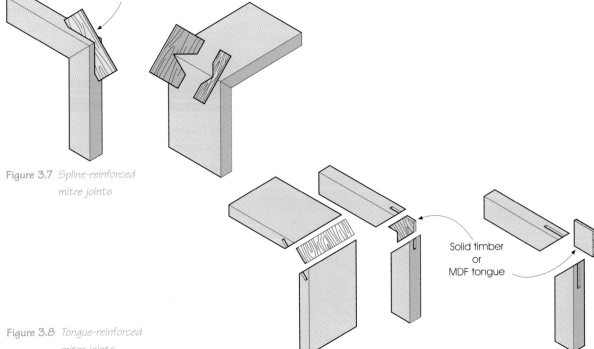

Veneer spline

Figure 3.7 *Spline-reinforced mitre joints*

Solid timber or MDF tongue

Figure 3.8 *Tongue-reinforced mitre joints*

Basic Woodworking Joints

Chapter 3

Forming a mitred butt joint

◆ Mark the cutting lines on the face or edge with a marking knife and mitre square. Square the lines onto the adjacent face or edge using a try square (Figure 3.6).

◆ Cut to the lines using a tenon saw. A mitre box may be useful to hold the workpiece and guide the saw.

◆ Trim the cut mitres with a plane and shooting board. Wide mitres for boxes can be trimmed and held in a vice, using an off-cut on the back edge to prevent breakout.

◆ Apply glue, nail or cramp joint together and remove excess glue with a damp cloth.

Spline reinforcement

This is carried out when the joint has been assembled. Use a tenon saw to cut slots across the corner. The cuts may be square to the end or angled like a 'dovetail' for additional strength. Glue thin plywood or veneer splines into the slots (Figure 3.7). Trim up flush when the glue has set.

Tongue reinforcement

The grooves or slots for this must be cut before the mitre is glued and assembled. Cut the tongue from 3 mm MDF or plywood. Solid wood may also be used but for strength make sure the grain runs across the width (Figure 3.8).

Cutting the groove for the tongue

◆ Use a mortise gauge to mark the parallel lines across the end grain.

◆ Cut down each line with a tenon saw.

◆ Chop out the waste with a chisel; alternatively, use a plough plane to form the groove.

◆ Centre the groove when the tongue is with the mitre, but move it off-centre, nearer to the inside of the mitre, when it runs across the mitre to minimise the weakening effect of short grain.

 Lap joints

In structural work the lap joint is used to lengthen timber. The two pieces overlap each other side by side, and are secured by nails, screws or bolts (Figure 3.9). Typical uses are for lengthening floor joists or pitched roof rafters. Laps must always be made over a support (wall plate or purlin).

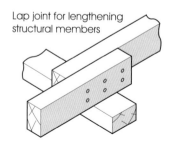

Lap joint for lengthening structural members

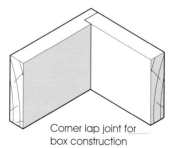

Corner lap joint for box construction

Figure 3.9 *Lap joints*

The lap joint is also a basic corner joint used for box and carcass construction.

It has advantages over a butt joint because of its increased gluing area, resistance given by the shoulder of inward pressure and a partial concealment of end grain. Reinforcement is given by nailing in one or both directions.

Forming a lap joint

◆ Cut the piece to length.
◆ Mark a shoulder line square across the face of the piece to be rebated and continue down both edges (Figure 3.10). The distance from the end will be the thickness of the other piece.
◆ Set a marking gauge to between a half to two-thirds the thickness of the rebated piece. Scribe a gauge line on the end grain and over each edge. Mark the waste with a pencil.
◆ Hold the rebate piece vertically in a vice. Use a tenon saw to cut down the gauge line to the shoulder line.
◆ Using a bench hook to hold the piece, cut across the shoulder line with a tenon saw to remove the waste. (Alternatively the shoulder line may be cut first and the waste removed by paring down the end grain with a chisel.)
◆ Apply glue to the rebated piece, assemble the parts and secure with pins in one or both directions. Remove excess glue with a damp cloth.

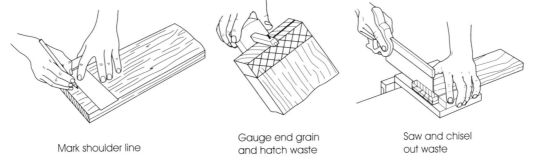

Mark shoulder line   Gauge end grain and hatch waste   Saw and chisel out waste

Figure 3.10 *Forming a lap joint – box construction*

Mitred lap joint

Also termed a splayed lap joint or scarf joint. It is used in structural work for lengthening in one line without increasing the cross-section at the joint. A long taper is cut on both pieces and are secured by nails, screws or bolts (Figure 3.11). Shoulders at right angles to the taper may be formed at either end for location.

A further variation is the wedged scarf joint, where the taper is stepped and the joint tightened with folding wedges.

The corner mitre lap joint is used for box construction. It is stronger than a plain mitre due to the locating shoulder and neater than the corner lap joint due to the absence of end grain.

Basic Woodworking Joints   **Chapter 3**

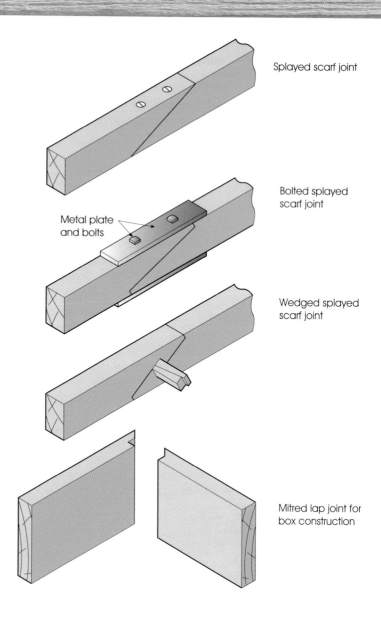

Splayed scarf joint

Bolted splayed
scarf joint

Metal plate
and bolts

Wedged splayed
scarf joint

Mitred lap joint for
box construction

Figure 3.11 *Mitred lap joints*

Forming a mitred lap joint

◆ Mark and cut the laps using a similar method to the corner lap, except that this time both pieces will be lapped (Figure 3.12).
◆ Mark a mitre on both pieces. Plane off the waste using a block plane for the rebated piece and a shoulder plane for the abutting piece.
◆ Glue and secure the joint as before.

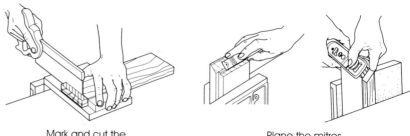

Figure 3.12 *Forming a mitred lap joint – box construction*

Mark and cut the
laps in both pieces

Plane the mitres

Halving joints

Halving joints are cut in pieces of equal thickness with half of the timber cut from each piece at the point of intersection. They are used for both lengthening and flat frame construction.

Scarf halving joint – are used for lengthening structural timber, typically wall plates where the joint is supported throughout its length (Figure 3.13). Nails or screws are used to secure the joint. An improvement is the bevelled scarf joints which will resist pulling stresses.

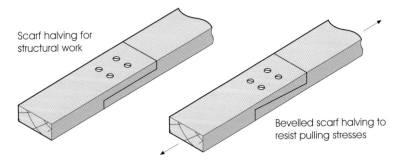

Scarf halving for structural work

Bevelled scarf halving to resist pulling stresses

Figure 3.13 *Scarf halving joints*

Angle halving joint – are used for both structural work and flat frame construction where one piece abuts or crosses over another (Figure 3.14).

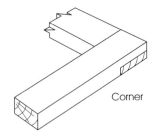
Corner

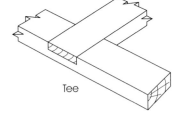
Tee

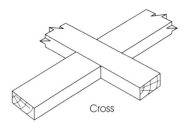
Cross

Figure 3.14 *Angle halving joints*

Forming a corner halving joint

- Use the piece of timber to mark the width on both face sides (Figure 3.15). Use a try square to mark the shoulder lines across the face and continue down the edges.
- Set a marking gauge to half the timber thickness and from the face side scribe a line on the edge from the shoulder, up to the end, across the end grain and finish up to the shoulder on the other edge. Mark the waste and repeat on the other piece of timber.
- Position one piece of timber vertically in the vice and saw a guide cut across the end grain on the waste side (half mark of the gauge line).
- Reposition the piece at an angle, keeping the saw in the guide cut. Make an angled cut down across the end grain and down to the shoulder line.
- Reverse the piece in the vice, position the saw in the guide cut and make a second angled cut down the other edge.
- Reposition the piece vertically in the vice and using the angled cuts as a guide, saw down to the shoulder line.
- Holding the piece on a bench hook, use a tenon saw to cut across the shoulder line and remove the waste.

Basic Woodworking Joints

Chapter 3

◆ Repeat the above processes to cut halving in the other piece.
◆ Dry assemble joint, then check for flatness and square. Clean up halving if required with a paring chisel.
◆ Glue up and cramp. Secure with pins or screws if required.

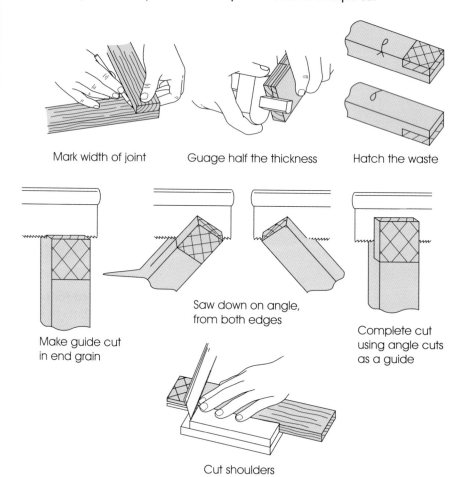

Mark width of joint   Guage half the thickness   Hatch the waste

Make guide cut in end grain

Saw down on angle, from both edges

Complete cut using angle cuts as a guide

Cut shoulders

Figure 3.15 *Forming a corner halving joint*

Mitred corner halving joint – is a refined version suitable for moulded timber trim (Figure 3.16). However, as the reduced gluing area makes this a weaker joint, it should be secured from the rear face with screws.

Single or double bevelled corner halving joint – can be used for structural work as they will resist pulling stresses in one or both directions (Figure 3.16). Marking out and cutting is the same as for standard halving except that the bevel is marked with a sliding bevel or template.

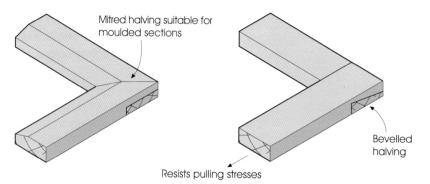

Mitred halving suitable for moulded sections

Bevelled halving

Resists pulling stresses

Figure 3.16 *Corner halving with moulded sections and with bevelled halving*

Forming a tee halving joint

- Mark and cut the upright member of the tee using the method described for the corner halving.
- Use the upright member to mark the width of the cut-out in the cross member of the tee.
- Use a try square to mark shoulder lines across the face and down the edges (Figure 3.17).
- With the previously set marking gauge, scribe a line on both edges between the marked shoulder lines. Mark the waste.
- Holding the piece with the aid of a bench hook, use a tenon saw to cut down the shoulders to the gauged line. Ensure you cut inside the waste. A tight joint can be eased, a slack one can't. A slack one is also unsightly and weak.
- Make one or more additional saw cuts across the waste to make chiselling out easier.
- Holding the piece in the vice, use a wide chisel and mallet to remove the waste. Work from both sides towards the centre with the chisel held at a slightly upward angle.
- Having removed most of the waste, use the chisel flat to pare away the raised centre portion. To cut any remaining fibres not cut by the saw, use the chisel vertically along the base of the shoulder line. Use the side edge of the chisel to check for flatness.
- Dry assemble joint, pare shoulders if too tight. Check for level and square. Adjust if required.
- Glue up and cramp. Secure with pins or screws if required.

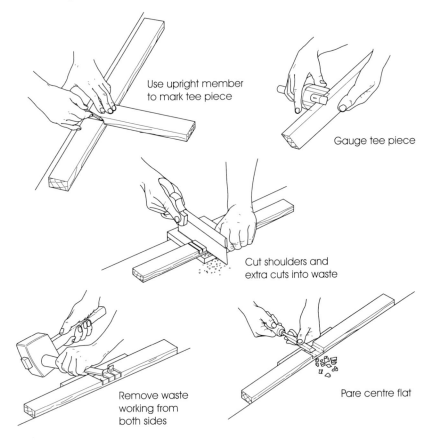

Use upright member to mark tee piece

Gauge tee piece

Cut shoulders and extra cuts into waste

Remove waste working from both sides

Pare centre flat

Figure 3.17 *Forming a tee halving joint*

Basic Woodworking Joints

Chapter 3

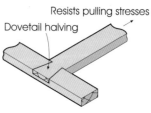

Resists pulling stresses

Dovetail halving

Figure 3.18 *Dovetail tee halving joint*

Dovetail tee halvings

These are an improvement on the standard joint as they resist pulling stresses better (Figure 3.18). The dovetail tee joint is formed in a similar way to the standard tee:

- Mark out and cut the piece to be inserted first, using the same method as the standard joint.
- Set a sliding bevel to the required slope and mark the pin (Figure 3.19).
- Saw and pare away the waste.
- Lay the dovetailed pin on the cross member.
- Mark the shoulder lines. Square the shoulder lines down the edges and score a gauge line between them.
- Saw shoulder lines, chisel out and assemble as described before.

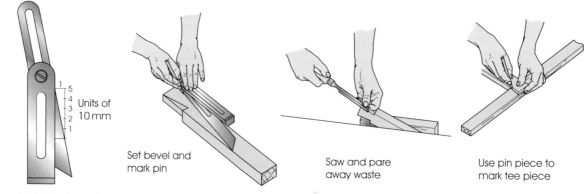

1 5 4 3 2 1 Units of 10 mm

Set bevel and mark pin

Saw and pare away waste

Use pin piece to mark tee piece

Figure 3.19 *Forming a dovetail tee halving joint*

Cross halving joints

These are used where horizontal rails cross vertical members, such as cabinet front frames and glazing bars of doors or windows. The joint in square timber is marked out and cut using a similar method to that described for the tee halving, except as both pieces are cross-members their shoulders should be cut with a tenon saw and the waste chiselled out.

Mitred cross-halving

These joints are used for moulded glazing bars. The basic joint is a cross-halving in the centre section with the moulding mitred (Figure 3.20).

- Mark the two central positions of the joint around both; a box square will be useful for this stage.
- Cut away the moulded section on both faces of each piece, down to the central section. The square cut of a mitre box can be used as a guide (Figure 3.21).
- Place a box mitre over the section and secure in the vice or with a cramp. Pare away the corners of each moulding in turn using the box mitre as a guide for the final cuts.
- Cut the cross-halving in the centre sections, making the cut-outs level with the rebate depth.

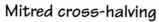
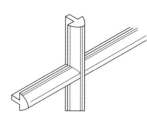
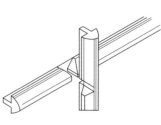

Figure 3.20 *Mitred cross-halving*

- Dry assemble, and pare mitres and shoulders as required if any are too tight. Take extra care when dry assembling and fitting as the slender remaining cross-section is easily snapped.
- Glue up and cramp; remove excess glue with a damp cloth.

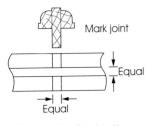
Mark joint

Equal

Equal

Make width of cut-out the same as width of central section

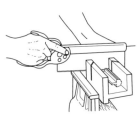
Cut away central section

Use box mitre as guide to pare waste

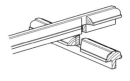
Carefully assemble joint

Figure 3.21 *Forming a mitred cross-halving joint*

Housing joints

Housings are grooves (also known as trenches) cut across the grain of a piece of timber to accommodate the square cut ends of another piece, thus forming a right-angled joint. They have a variety of uses ranging from stud partitions and stairs to fixed shelving and intermediate standards in cabinet carcasses.

Through housings

The depth of housings should be restricted to about one-third the material thickness (Figure 3.22).

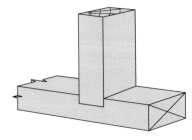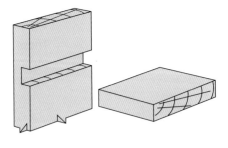

Figure 3.22 *Through housing joint*

Forming a through housing joint

- Measure and mark in pencil the position of the groove (Figure 3.23).
- Mark square parallel lines across the piece, equal to the thickness of the stud or shelf, etc. Continue the line down the edges.
- Score shoulder lines across the face, using a marking knife and try square.
- Set the marking gauge to one-third the thickness and scribe lines between the marks on both edges.

- Cut down on the waste sides of the shoulder lines to the gauge lines. On wide housings make further saw cuts in the waste to ease chiselling out the waste.
- Chisel out the waste working from both sides towards the centre. This will help to prevent the risk of breakout.
- Pare the bottom of the housing flat. Alternatively, for housings in wide boards, use a hand router plane to remove the waste between the sawn shoulder lines. Start off making shallow cuts, resetting the cutter deeper each time until the full depth is reached.
- Dry assemble, easing if required, before gluing up.

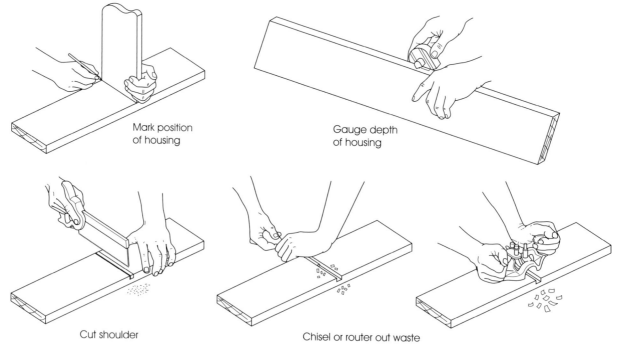

Mark position of housing

Gauge depth of housing

Cut shoulder

Chisel or router out waste

Figure 3.23 *Forming a through housing*

Dovetail housings

These are a variation of the through housing and are used to resist endwise pulling stresses. The housing may be cut with one or both sides of the groove cut at an angle (Figure 3.24).

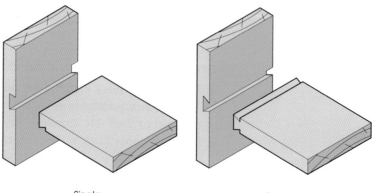

Figure 3.24 *Dovetail through housing joint*

Single

Double

Forming a dovetail housing joint

The marking and cutting of the dovetail housing is similar to the standard through housing, except that the lines down the edges are marked using the dovetail outline of the shelf, with the shoulder lines cut to the marked angles. An angled block may be cramped along the shoulder line to act as a guide for the tenon saw.

- Mark the shoulder line on the tailpiece (end of shelf) equal to the depth of housing and score with a marking knife. Square the shoulder line around the edges.
- Mark the dovetail shape from the end of the shelf towards the shoulder line using a sliding bevel set to a suitable dovetail angle.
- Saw along the shoulder line down to the marked angle.
- Pare out the waste using a purpose-made block to guide the chisel (Figure 3.25).
- Mark out the shelf housing, using the shelf as a guide. Square the shoulder lines across the face.
- Saw along the shoulder lines using an angled block to guide the saw.
- Remove the waste using a bevel edged chisel.
- Dry assemble, easing the shoulders and bottom of the groove if required.

The joint is assembled by sliding the tailpiece into the housing from one end.

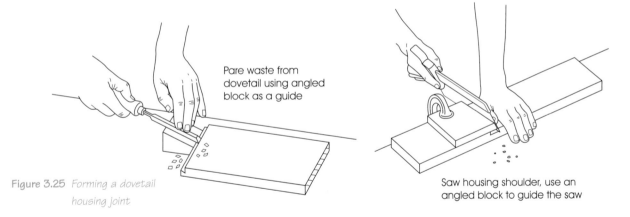

Pare waste from dovetail using angled block as a guide

Saw housing shoulder, use an angled block to guide the saw

Figure 3.25 *Forming a dovetail housing joint*

Stopped housing joints

The groove of a stopped housing is stopped short of the full width of the timber (Figure 3.26). For use where the member to be joined is narrower, as in set back, fixed shelves and stair strings.

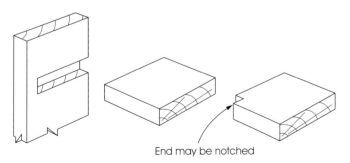

Figure 3.26 *Stopped housing joint*

End may be notched

Basic Woodworking Joints

Chapter 3

Forming a stopped housing joint

◆ Mark out the groove partially across the face and down one edge, as described for the standard joint. Gauge the stopped end of the groove.

◆ Bore a hole in the stopped end of the groove down to the required depth using a flat bottomed Forstner bit (Figure 3.27).

◆ Square up the hole using a chisel. Alternatively chop a cut-out at the stopped end using a chisel and mallet. Wrap a piece of masking tape around the chisel blade to indicate the required depth.

◆ Starting from the cut-out at the stopped end, use the tip of a tenon saw to cut along the shoulder lines, down to the required depth.

◆ Remove waste from the groove with a chisel and/or a hand router.

◆ Dry assemble, easing the shoulders and bottom of the groove if necessary.

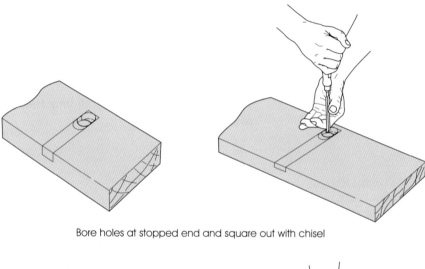

Bore holes at stopped end and square out with chisel

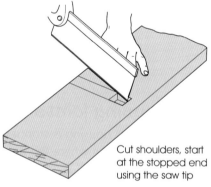

Cut shoulders, start at the stopped end using the saw tip

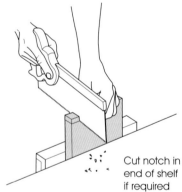

Cut notch in end of shelf if required

Figure 3.27 *Forming a stopped housing joint*

Stopped housings can also be cut as dovetails where resistance to endwise pull-out is required. A further variation can be made to the stopped housing which conceals the end of the groove and masks the effects of subsequent shrinkage. This involves cutting a shoulder or notch on the front edge of the shelf.

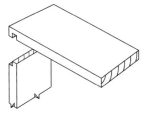

Figure 3.28 *Bare-faced housing joint*

Bare-faced housing joint

These are also termed bare-faced tongue joints. It is a stronger corner joint than the standard housing (Figure 3.28) and often used for joining door linings. The groove is cut in the head (horizontal member) and the one shouldered tongue on the end of the jamb (vertical member).

Forming a bare-faced housing joint

◆ Mark the width of the vertical member on the face of the horizontal member, square across the face and onto the edges (Figure 3.29).
◆ Set the marking gauge to one-third the thickness. Use the gauge to mark the shoulder line across the back face and edges of the vertical piece. Then also across the end grain from the face side to mark the width of the tongue.
◆ Continue the lines on either edge to meet the shoulder lines.
◆ Saw down the shoulder line.
◆ Cut out the waste with a chisel, by paring the end grain.
◆ Align the face side of the tongue with the line marked on the horizontal piece. Use the tongue to mark the width of the housing. Square the lines across the face and onto the edges.
◆ Use the pre-set gauge to mark the depth of the housing on the edges.
◆ Saw and chisel out the housing as before.
◆ Dry assemble, easing if required before gluing up.

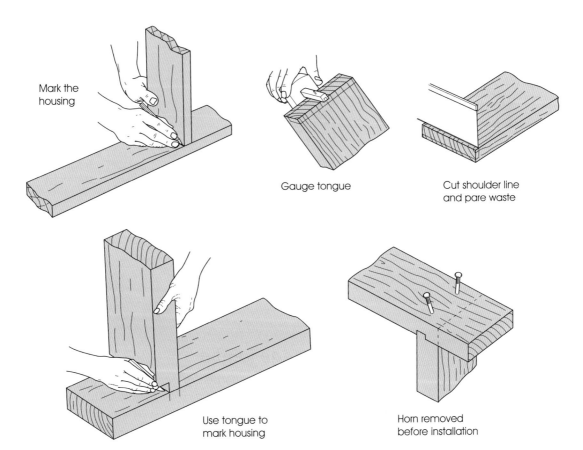

Mark the housing

Gauge tongue

Cut shoulder line and pare waste

Use tongue to mark housing

Horn removed before installation

Figure 3.29 *Forming a bare-faced housing joint*

Basic Woodworking Joints

Chapter 3

This joint is often reinforced by nailing or screwing. It is normal practice to leave the horizontal member long (leaving a horn) when working and assembling the joint. The horn, which helps to prevent grain splitting, can be cut off before the assembled component is installed.

Edge joints

Edge joints are used to enable narrow boards to be built up to cover large areas for floor boarding or cladding or to form wider boards for shelves, cabinet carcasses, panels, counter/worktops and table tops, etc.

The edges of the boards may be plain (butt) or shaped (Figure 3.30). The shaping provides a means of interlocking to line up the surfaces. For glued joints it increases the gluing area for additional reinforcement.

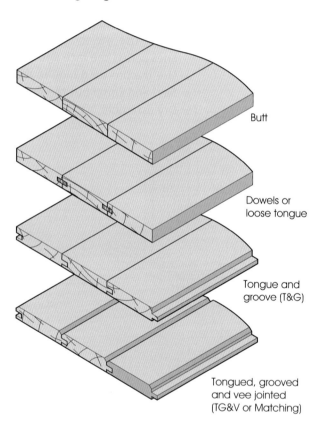

Butt

Dowels or loose tongue

Tongue and groove (T&G)

Tongued, grooved and vee jointed (TG&V or Matching)

Figure 3.30 Edge joints

Unglued edge joints

These are mainly used in carpentry work for flooring and cladding. Each board will expand or contract across their width with changes in the moisture content. On shrinking, gaps will appear between the boards. Narrow boards will show a smaller gap on shrinkage than wide boards. Tongued and grooved (T&G) edge joints are normally used for flooring. The interlocking increases the strength and prevents a through gap on shrinkage.

Glued edge joints

These are used to join narrow boards together forming a wide panel for use in cabinet work. The joints may be simply butted or reinforced with a tongue or dowels, to ease location and increase the glue line.

Preparing the edges

◆ Lay out boards to be joined, selecting them for colour and grain direction. When using tangential sawn timber, make sure the heart side alternates to minimise the effect of distortion.

◆ Number each board and mark face and edge marks (Figure 3.31). When working with them keep numbers facing the same way.

◆ Set the first two boards 1 and 2 back to back in the vice and use a try plane to true the edges. Using this method, the squareness of the edge is not critical, as they will still fit together and form a flat surface.

◆ Repeat the previous stage using boards 2 and 3 again back to back. Note that board 2 will have to be rotated in order to match its unplaned edge with board 3. Continue to plane the edges of each pair of boards.

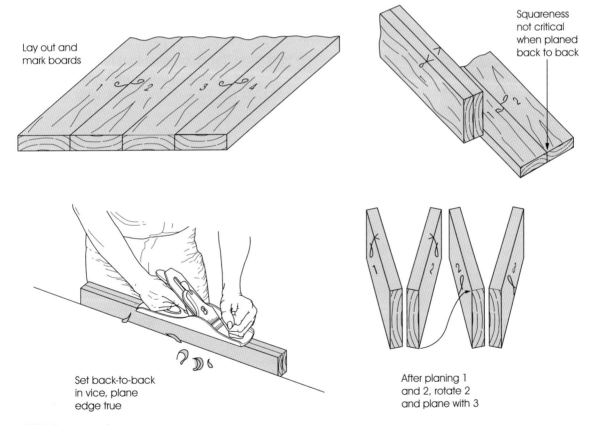

Lay out and mark boards

Squareness not critical when planed back to back

Set back-to-back in vice, plane edge true

After planing 1 and 2, rotate 2 and plane with 3

Figure 3.31 *Preparing edge joints*

Gluing and cramping the joint
Rubbed joint

A close fitting butt joint can often be glued up without any applied pressure (cramping).

◆ Apply glue to both edges.

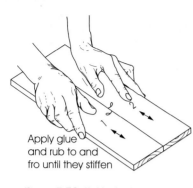

Apply glue
and rub to and
fro until they stiffen

Figure 3.32 *Rubbed edge jointing method*

◆ Bring the edges together and rub to and fro until they start to stiffen (Figure 3.32). This action squeezes out the excess glue and air from the joint and brings the components into close contact.

◆ Remove excess glue with a damp cloth and allow to cure before truing up joined faces with a smoothing plane.

Cramped joint

Where more than two boards are being jointed, the use of cramps is preferable to a rubbed joint.

◆ Lay out the numbered boards on the bench. Use battens at right angles for support.

◆ Dry assemble and check the boards for fit. Use two or more sash cramps, positioned about a quarter of the boards' length from each end (Figure 3.33). Place scrap pieces between the cramps and the boards.

◆ If all joints are close fitting remove the cramps. Apply glue to the jointing edges and re-cramp.

◆ If necessary, tap any misaligning joints with a hammer and block of wood to make them flush.

◆ Turn the panel over and cramp the centre. This centre cramp will pull up the joints and also helps to keep the panel flat.

◆ Use a damp cloth to remove the excess glue that has squeezed out.

◆ Leave the panel in the cramps until the glue has cured, before finally cleaning up the face of the panel with a smoothing plane.

did you know?

Do not forget to use scrap pieces between the cramp heads and the board edges, to prevent damage.

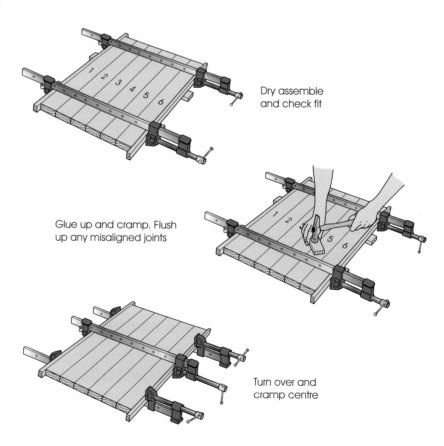

Dry assemble
and check fit

Glue up and cramp. Flush
up any misaligned joints

Turn over and
cramp centre

Figure 3.33 *Cramping edge joints*

Notched and cogged joints

These joints are used in heavy carpentry work to provide a location and ensure a uniform depth. Single joints cut in on one member allow movement in one direction (Figure 3.34). Double joints are cut in both members to prevent movement in both directions. The method used to form notched and cogged joints is similar to that used for housing and halving joints, except that the joint proportions and the timber section are normally much larger.

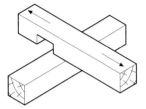

Single notched
locates in one direction

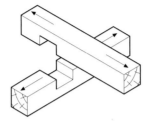

Double notched
locates in both directions

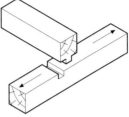

Single cogged
locates in one direction

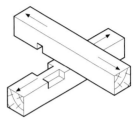

Double cogged
locates in both directions

Figure 3.34 *Notched and cogged joints for structured work*

Dovetail joints

Dovetail joints are designed to resist tensile or pulling forces. They are used mainly in box and drawer construction. Through dovetails are used for boxes and backs of drawers. On the front of drawers lapped dovetails are used because they provide a neater finished appearance (Figure 3.35).

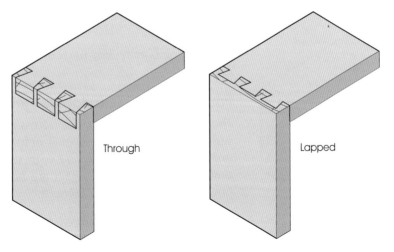

Through   Lapped

Figure 3.35 *Dovetail joints*

Dovetail joints should have a pitch or slope of one in six for softwood and one in eight for hardwood (Figure 3.36). An excessive slope is weak due to the short grain, whereas an insufficient slope will tend to pull out under load. The slope can be marked out using a dovetail template or a sliding bevel.

Basic Woodworking Joints **Chapter 3**

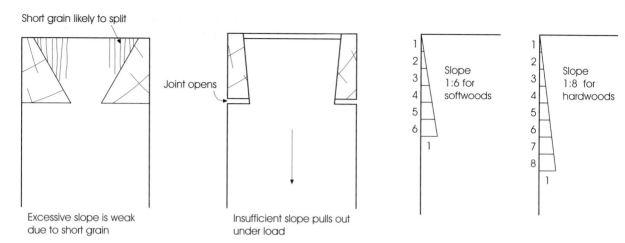

Short grain likely to split

Joint opens

Slope 1:6 for softwoods

Slope 1:8 for hardwoods

Excessive slope is weak due to short grain

Insufficient slope pulls out under load

Figure 3.36 *Dovetail angles*

Two methods (Figure 3.37) may be adopted when marking out and cutting dovetails, either:

◆ mark and cut the dovetail and use the dovetail to mark out the pins and sockets, or
◆ mark and cut the pins and sockets, then from these mark out the dovetails.

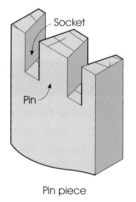
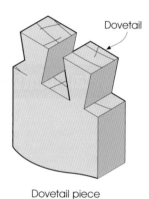

Socket

Dovetail

Pin

Pin piece

Dovetail piece

Figure 3.37 *Dovetails, pins and sockets*

Dovetail joints may also be formed using a powered router and jig (see Chapter 4).

Forming a through dovetail joint

The following operations are necessary when making a dovetail joint in wide boards using the dovetail before pin method. The actual size and number of dovetails will vary with the width of the board and type of timber. In softwoods, dovetails are normally cut wider and thus fewer in number than those in hardwoods.

◆ Cut the ends of the pieces to be joined square. The end grain can be trued up and smoothed with the aid of a shooting board.
◆ Use a pencil and try square to mark the thickness of the board around the ends of both pieces.
◆ Divide the width of the board by the number of dovetails (Figure 3.38).

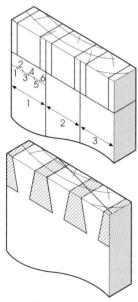

Figure 3.38 *Marking out dovetails*

- ◆ Sub-divide each division into six equal parts, one at either end for the pin and four in the middle for the dovetail.
- ◆ Square lines over the end grain and down to the shoulder line.
- ◆ Use a dovetail template or a sliding bevel and pencil to mark the slope of the dovetails on both faces (Figure 3.39).

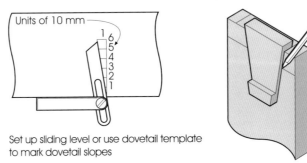

Units of 10 mm

Set up sliding level or use dovetail template to mark dovetail slopes

Figure 3.39 *Using bevel or dovetail template*

- ◆ Hatch out the waste with a pencil to avoid later confusion.
- ◆ Position the piece of wood in the vice at an angle, so that one slope of each dovetail is vertical (Figure 3.40).

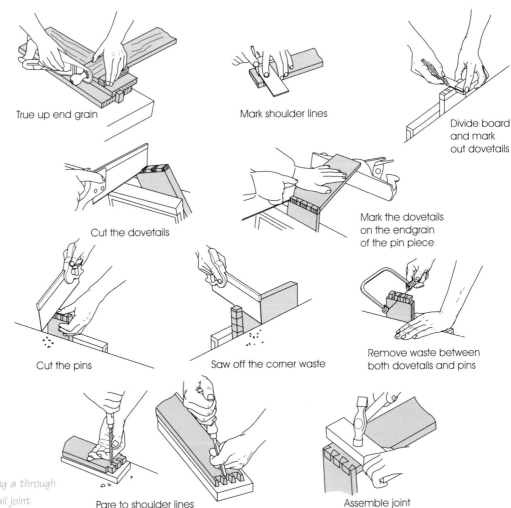

True up end grain

Mark shoulder lines

Divide board and mark out dovetails

Cut the dovetails

Mark the dovetails on the endgrain of the pin piece

Cut the pins

Saw off the corner waste

Remove waste between both dovetails and pins

Pare to shoulder lines

Assemble joint

Figure 3.40 *Forming a through dovetail joint*

- ◆ Use a dovetail saw to cut down one side of each dovetail, keeping just in the waste.

◆ Reposition the piece in the vice and cut down the other slope of each dovetail.

◆ Position the pin piece of wood vertically in the vice. Lay the cut dovetail piece in position, lining up the edges and shoulder line. Use a dovetail saw in each dovetail saw-cut to mark their slope on the end grain. Square these lines down to the shoulder line on both faces. Again, hatch the waste.

◆ With the pin piece positioned vertically in the vice, use a dovetail saw to cut down to the shoulder line, following the angle marked from each dovetail. Keep to the waste side, aiming to leave it just visible.

◆ Saw the corner waste from the dovetail piece.

◆ Use a coping saw to remove most of the waste between both the dovetails and the pins.

◆ Working from both faces, use a bevel edged chisel to pare out the remaining waste down to the shoulder lines.

◆ Partly assemble the joint to check the fit, paring out any high spots if required.

◆ Clean up the inside faces of both pieces.

◆ Apply glue to both halves of the joint and tap the joint together. Use a piece of waste wood to protect the surface.

◆ Use a damp cloth to remove the excess glue.

◆ Allow the glue to set before cleaning up the end grain with a smoothing or block plane. Work in from both edges towards the centre to avoid the end grain breaking out.

did you know?

Glue and sawdust is no substitute for tight-fitting joints.

Forming a lapped dovetail joint

This joint is formed using a similar process to that of the through dovetails, except for cutting the pin piece:

◆ Cut and prepare the ends of the pieces to be joined (Figure 3.41).

◆ Use a pencil and try square to mark two-thirds the thickness of the pin piece around the end of the dovetail piece.

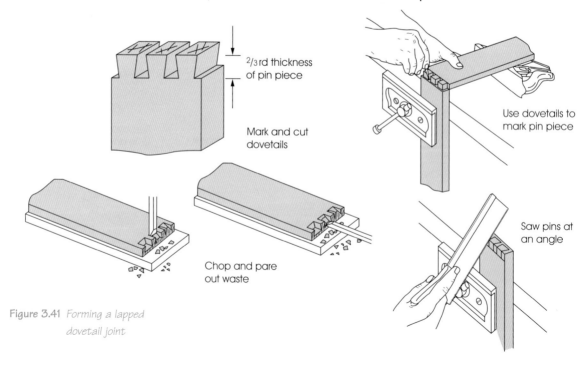

2/3 rd thickness of pin piece

Mark and cut dovetails

Use dovetails to mark pin piece

Chop and pare out waste

Saw pins at an angle

Figure 3.41 *Forming a lapped dovetail joint*

- Set out the dovetails and hatch the waste as before.
- Cut dovetails with a dovetail saw, using a coping saw to remove most of the waste. Pare to the shoulder line with a bevel edge chisel.
- Use a pencil and try square to mark the thickness of the dovetail piece on the inside face of the pin piece.
- Lay the dovetail piece on the pin piece; line up the edges and shoulder line. Pencil around the dovetails to mark the pins.
- Square the pin lines down to the shoulder line and hatch the waste.
- Position the pin piece vertically in the vice. Use a dovetail saw, held at an angle, to cut down the waste, stopping at the lap and shoulder lines.
- Use a cramp to secure the pin piece to the bench top.
- Chop out the waste using a chisel and mallet. Start just in from the end and work back towards the shoulder.
- From the end of the pin piece, pare out the remaining waste and trim into the corners using a bevel edged chisel.
- Clean up, then fit and assemble the joint as above.

Mortise and tenon joints

This joint is probably the most widely used angle joint. It is used extensively for door and window construction and general framework in a variety of forms (Figure 3.42).

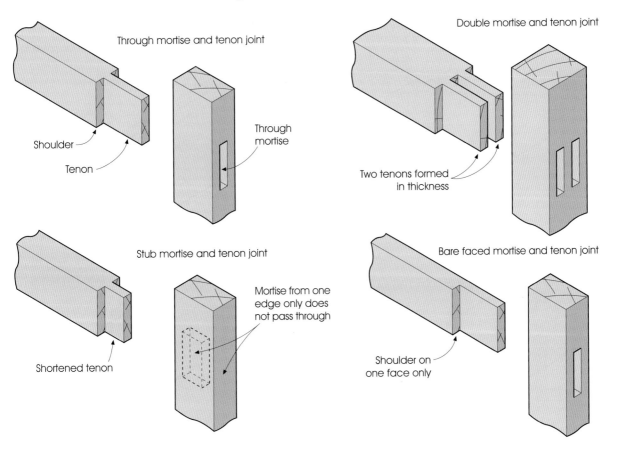

Through mortise and tenon joint

Shoulder

Tenon

Through mortise

Double mortise and tenon joint

Two tenons formed in thickness

Stub mortise and tenon joint

Shortened tenon

Mortise from one edge only does not pass through

Bare faced mortise and tenon joint

Shoulder on one face only

Basic Woodworking Joints

Chapter 3

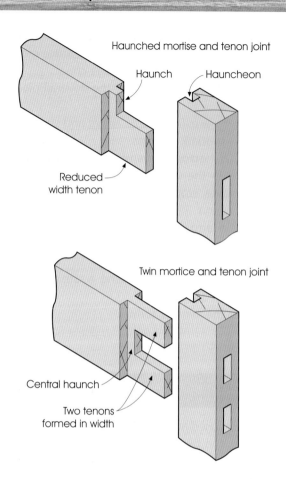

Haunched mortise and tenon joint

Haunch

Hauncheon

Reduced width tenon

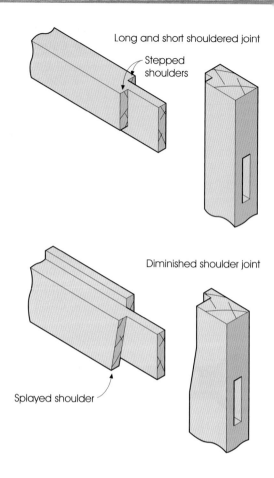

Long and short shouldered joint

Stepped shoulders

Twin mortice and tenon joint

Central haunch

Two tenons formed in width

Diminished shoulder joint

Splayed shoulder

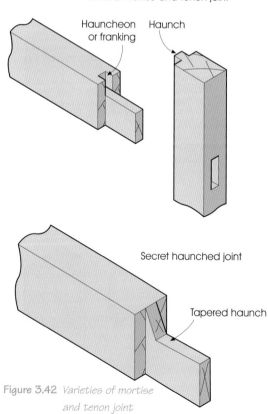

Franked mortise and tenon joint

Hauncheon or franking

Haunch

Secret haunched joint

Tapered haunch

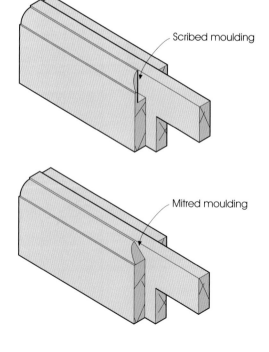

Moulded frame mortise and tenon joints

Scribed moulding

Mitred moulding

Figure 3.42 *Varieties of mortise and tenon joint*

Through mortise and tenon joint – where the rectangular mortise slot goes right through the timber member.

Stub mortise and tenon joint – where the mortise slot is stopped short and does not pass completely through the timber member. Also termed a blind mortise.

Haunched mortise and tenon joint – where the tenon has been reduced in width leaving a shortened portion of tenon protruding called a haunch. The purpose of forming a haunch rather than cutting the reduced part of the tenon flush with the shoulder is to provide a positive location for the full width of the rail, preventing it from twisting. Haunches are used in wide members to reduce the width of the tenon or where the joint is at the end of a framed member to permit wedging up.

Twin mortise and tenon joint – where a haunch is formed in the middle of a wide member, creating two tenons, one above the other.

Double mortise and tenon joint – where two tenons are formed in the thickness of a member, side by side.

Bare faced mortise and tenon joint – where the tenon has a shoulder on one face only. Used for stair strings and framed match-boarded doors.

Long and short shouldered joint – where the shoulder lines are stepped. Used for joints in rebated members.

Diminished shoulder joint – where one or both of the shoulders are splayed to accommodate a change in section above and below the mortise and tenon joint.

Franked mortise and tenon joint – where the haunch is formed on the mortised member and franking or hauncheon is cut into the tenoned member; used where a standard haunch would remove too much timber and weaken the joint. Also known as a sash haunch.

Secret haunched joints – also known as a table haunch, has a tapered haunch which does not show on the end grain.

Moulded frame mortise and tenon joint – where the members have moulded edges it is necessary to scribe or mitre the moulding. It is normal to make the depth of the moulding the same as the rebate, so that the shoulders are not stepped.

Joint proportions

The proportions of the mortise and tenon are important to the strength of the joint (Figure 3.43).

- ◆ The tenon should be one-third the thickness of the timber to be joined. If a chisel this size is not available to chop the mortise to receive the tenon, the thickness of the tenon may be adjusted to the nearest available chisel size.
- ◆ The width of the tenon should not exceed five times its thickness. This is to overcome the tendency of a wide tenon to buckle and also to reduce the effect of shrinkage.

did you know?

The sinking adjacent to the mortise to receive a haunch is termed a hauncheon.

◆ Where haunches are used to reduce the tenon width, they should be about one-third the width of the member and equal in length to the thickness of the tenon. Haunches between twin tenons are one-third of the member for middle rails or one quarter of the member where twin haunches are used for bottom rails.

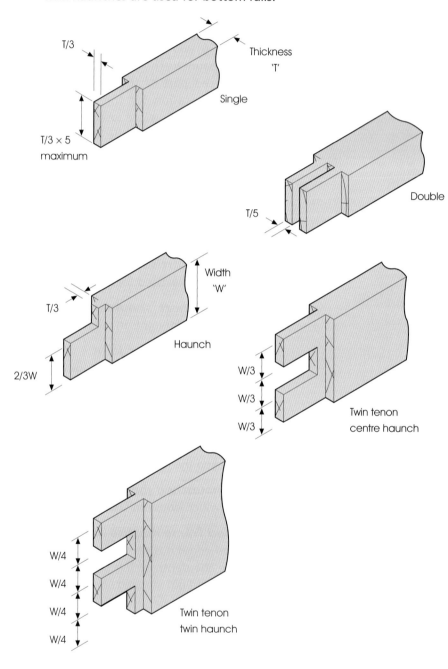

Figure 3.43 *Joint proportions*

Forming a through mortise and tenon joint

◆ Use a square and pencil to mark the shoulder line for the tenon, around the member (Figure 3.44). Then score the shoulder lines with a marking knife. Where tenons are required at either end, mark out the distance between the shoulder precisely.

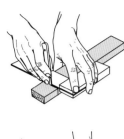

1. Mark shoulder lines

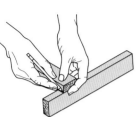

2. Mark mortise position

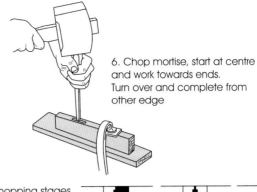

6. Chop mortise, start at centre and work towards ends. Turn over and complete from other edge

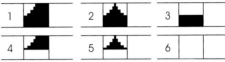

Chopping stages – top side	1	2	3

Chopping stages – reverse side	4	5	6

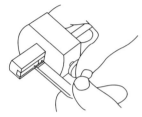

3. Set gauge to width of chisel

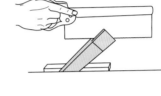

7. Saw diagonally down tenon lines

4. Gauge mortise

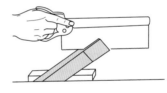

8. Reset and saw down tenon lines from other edge

5. Gauge tenon

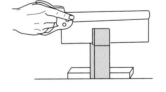

9. Saw down to shoulder using diagonal cuts as a guide

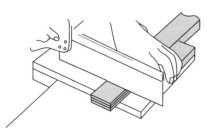

10. Saw shoulder lines

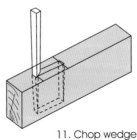

11. Chop wedge tapers

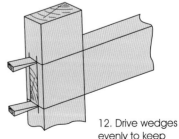

12. Drive wedges evenly to keep member in right position

Figure 3.44 *Forming a through mortise and tenon joint*

Basic Woodworking Joints

Chapter 3

- Mark the edge of the mortise on the mortise member and use the tenon member to mark its width.
- Square lines all around the member using a pencil.
- Set the pins of a marking gauge to the selected mortise chisel. Set the stock of the gauge to centre the mortise on the edge of the member.
- Using the gauge stock from the face side, score lines on both edges of the mortise member and from the tenon shoulder line on one edge, over the end and back to the shoulder line on the other edge.
- Cramp the mortise member to the bench or secure it in the vice. Position the mortise chisel in the middle of the mortise, holding it vertically with the cutting edge at right angles to the gauged lines.
- Drive the chisel with a mallet to the depth of 3 or 4 mm. Work backwards towards the end of the mortise in 3 or 4 mm steps. Each time the chisel is driven it will cut progressively deeper so that when the mortise end is reached it will be about halfway through. Use a to-and-fro rocking action to release the chisel and break the waste.
- Turn the chisel around and work from the centre, back towards the other end.
- Turn the member over, tap out any loose waste before securing to avoid bruising the work. Chop out the waste from the other edge using the same process, until the mortise is cut through.
- Secure the tenon member at an angle in a vice. Saw diagonally down each tenon line and across the end grain, keeping just to the waste side. Reset in the vice and saw down the tenon lines from the other edge. Place the member upright in the vice and saw down level to the shoulder using the diagonal saw cuts as a guide.
- Hold the member on a bench hook and use a tenon saw along the shoulder lines to remove the waste. Take care not to cut too deep and weaken the tenon.
- Dry assemble the joint. The tenon should fit straight from the saw, but may be pared with a chisel if it's too tight. Joints that do not pull up, may be eased by working a tenon saw along the shoulder on both sides, with the joint assembled. Again, take care not to cut into the tenon.
- Working from the outside edge, make a tapered cut with the chisel about 6 mm wide at either end of the mortise. Cut two wedges from a piece of timber the same thickness as the tenon.
- Clean up the inside edges before applying the glue to the tenon and the shoulders. Assemble the joint and cramp if required.
- Apply the glue to the wedges and tap them in, striking each alternately to keep the members in the right location and the wedges level.
- Once the glue has cured, clean up the faces with a smoothing plane.
- Saw off any protruding ends of the tenon and wedges and finish by planing the ends flush.

Wedging stub tenons

This is normally done using fox-wedges.

- Chisel out a tapered undercut to both ends of the mortise. Cut two wedges the same thickness as the tenon, about two-thirds of the tenon length and 3–4 mm from the widest point of the taper (Figure 3.45).

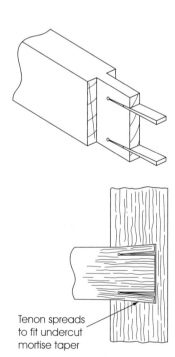

Tenon spreads
to fit undercut
mortise taper

Figure 3.45 *Fox wedges*

- Make two saw cuts in the tenon about 6mm in from either edge and just longer than the wedges.
- Drill a small hole at the end of each cut, to help prevent splitting the tenon.
- Clean up the inside edges and apply glue to the tenon, shoulders and wedges.
- Start the wedges in the saw cuts and assemble the joint. When cramping, the wedges will be driven into the cuts by the bottom of the mortise, causing the tenon to spread and fit tightly against the undercut mortise tapers.
- Clean up the faces once the glue has cured.

Haunches

These are cut out after the shoulders have been removed.

- Mark the tenon width from the previously cut mortise (Figure 3.46).

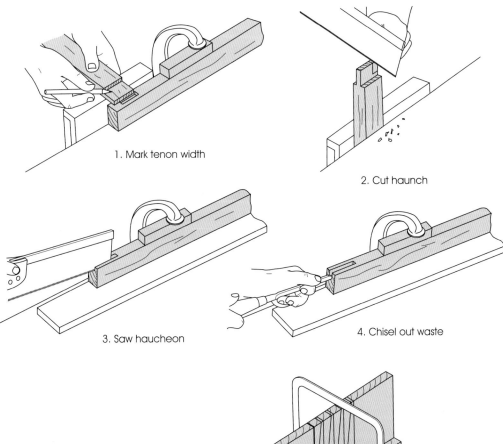

1. Mark tenon width

2. Cut haunch

3. Saw haucheon

4. Chisel out waste

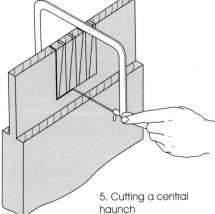

5. Cutting a central haunch

Figure 3.46 Forming haunches

Basic Woodworking Joints

Chapter 3

When making haunched
mortise and tenon joints, the
waste cut from the haunch
may be used for cutting the
wedges.

◆ Mark the length of the haunch and cut along both lines with a tenon or panel saw.
◆ Cut the hauncheon with a tenon saw and chisel out the waste. A coping saw may be used to cut out the waste from central haunches between twin tenons.

The part of the tenon cut away to form the haunch may be used for cutting wedges.

Bridle joints

These joints are a form of mortise and tenon, traditionally used for jointing heavy timber frames and roof trusses (Figure 3.47). They are now used for light framing in joinery and cabinetwork. They are also known as forked mortise and tenon joints when used in the middle of a member.

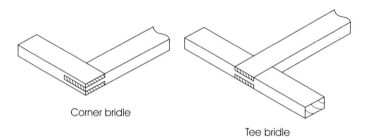

Corner bridle

Tee bridle

Figure 3.47 *Bridle joints*

Dowel joints

Dowels provide a simple method of joining solid timber framework, instead of using mortise and tenons, and joining sheet material for the carcasses of cabinets, etc. They are in effect a butt joint that has been reinforced with small wood pegs, glued into holes drilled into both parts being jointed (Figure 3.48).

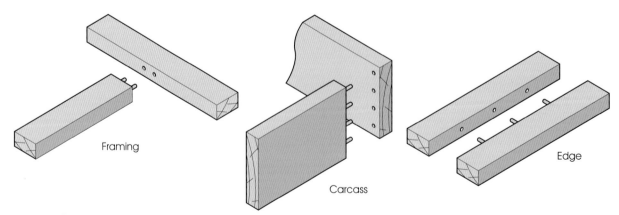

Framing

Carcass

Edge

Figure 3.48 *Dowel joints*

Dowels themselves are made from hardwood. They may be cut from a length of commercially produced dowel rod. Alternatively 'ready made' chamfered-end, fluted dowels are available (Figure 3.49).

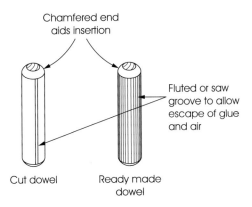

Figure 3.49 *Dowels*

Use a dowel diameter that is up to half the thickness of the pieces being joined and a minimum length of five times its diameter. Longer dowels give a greater gluing area and thus form a stronger joint.

When making your dowels they should be prepared by:

- sawing to length;
- chamfering the ends to aid location and insertion;
- cutting a groove down the length of each dowel to allow excess glue and trapped air to escape when the joint is assembled.

Marking out dowel joints

Accurate positioning and drilling of dowel holes in each piece is essential. These may be marked out using a number of methods.

Edge-to-edge and framing joints – may be marked out, by placing the two pieces flush in the vice, with the face sides out (Figure 3.50). Square pencil lines across the edge at each dowel position. Set a gauge to half the timber thickness and score the centre line on each, working from both face sides. The number and spacing of dowels depends on size: at least three for edge joints 100 to 150 mm apart; at least two for framing joints; three for rails over 100 mm wide. Centre the drill where the lines cross and bore the holes. These should be a little deeper than half the dowel length, to ensure a tight fitting joint and to create a glue reservoir.

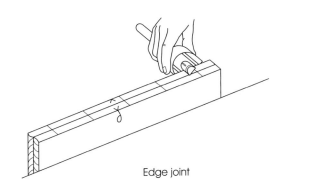

Edge joint

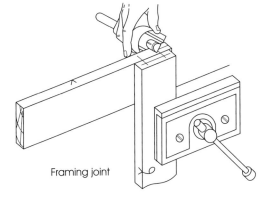

Framing joint

Figure 3.50 *Marking out for dowels*

Centre pins – may be used to mark one part from the other (Figure 3.51). You can use panel pins or buy some proprietary dowel centre points. Using the panel pin method, mark the dowel positions on either edge of an edge-to-edge joint, or on the end of the rail for framing joints. Drive panel pins into the centre marks and crop off the heads with a pair of pincers, leaving about 6mm protruding. Lay both members flat on the bench and push together to transfer the centre points.

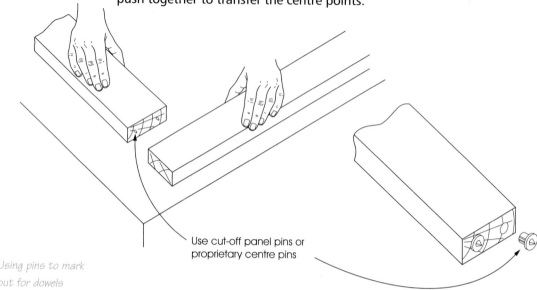

Use cut-off panel pins or proprietary centre pins

Figure 3.51 *Using pins to mark out for dowels*

Using proprietary centre pins, mark and drill the holes in one edge or end of the rail. Insert the dowel pins into each hole and bring together as before.

Templates – may be made to aid marking out (Figure 3.52). Cut a thin piece of plywood or MDF the same size as the cross-section of the timber or carcass panel. Mark the dowel centres on the template. Position over the joint and drive panel pins through to mark the centres in each piece.

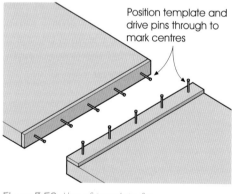

Position template and drive pins through to mark centres

Figure 3.52 *Use of template for marking dowels*

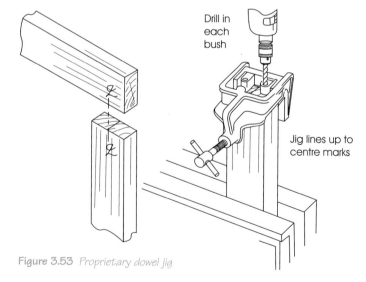

Drill in each bush

Jig lines up to centre marks

Figure 3.53 *Proprietary dowel jig*

Proprietary dowelling jigs – cramp onto the workpiece, position the holes and guide the drill accurately (Figure 3.53). Most have interchangeable guide bushes for 6, 8 and 10mm drills and dowels.

Assembling frames

When making frames, each joint should be cut separately and the entire frame dry assembled to check the fit of the joints, overall sizes, square and winding, before gluing and cramping up.

Squaring up

A frame is checked with a squaring rod, which consists of a length of rectangular section timber with a panel pin in its end (Figure 3.54). The end with a panel pin is placed in one corner of the frame. The length of the diagonal should then be marked in pencil on the rod. The other diagonal should then be checked. If the pencil marks occur in the same place, the frame must be square. If the frame is not square, then sash cramps should be angled to pull the frame square.

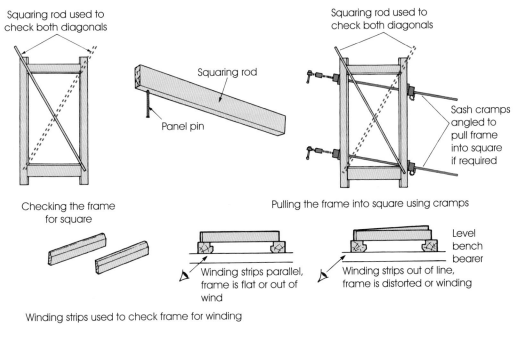

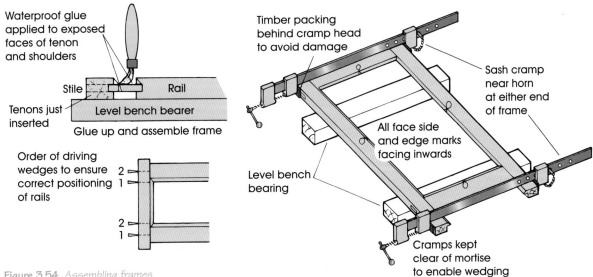

Figure 3.54 Assembling frames

Winding

A frame is checked with winding strips. These are two parallel pieces of timber. With the frame laying flat on a level bench, place a winding strip at either end of the job. Close one eye and sight the tops of the two strips. If they appear parallel the frame is flat or out of wind. The frame is said to be winding, in wind or distorted if the two strips do not line up. Repositioning in the cramps or adjustment to the joints may be required.

Glue up

Assemble and lightly drive wedges. A waterproof adhesive should be used for external joinery or where it is likely to be used in a damp location. Ensure the overall sizes are within the stated tolerances. Re-check for square and wind. Assuming all is correct, finally drive the wedges.

Board fittings, edging and laminating

KD fittings

Most joints used for joining solid wood can also be used for joining board material. However, board material is more commonly joined using butt joints, either glued and screwed together or reinforced with dowels or biscuits; or secured with a proprietary knock down (KD) fitting (Figure 3.55).

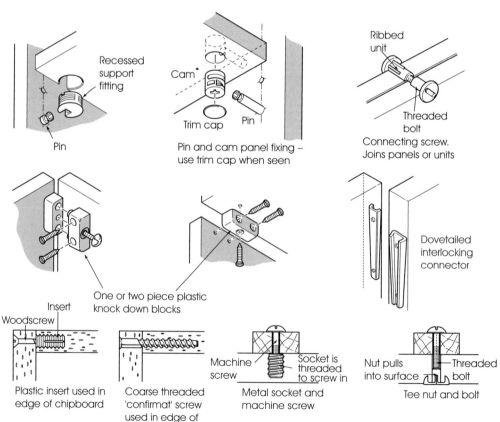

Figure 3.55 *Knock down fittings*

Edging board material

Edging is also termed lippings. They are applied to the exposed edges of board material to cover the core (Figure 3.56).

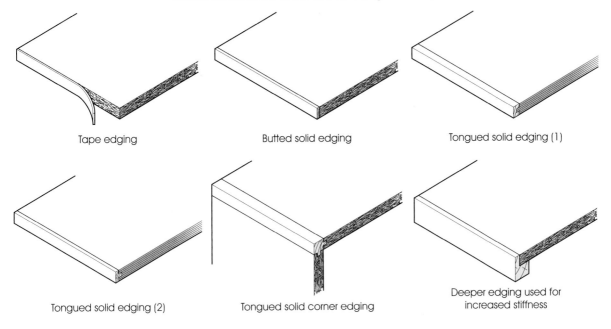

Tape edging

Butted solid edging

Tongued solid edging (1)

Tongued solid edging (2)

Tongued solid corner edging

Deeper edging used for increased stiffness

Figure 3.56 *Edging board material*

Pre-glued iron-on tape

This is the simplest edging either available in a range of veneers or melamine colours to match the board finish (Figure 3.57). Use brown paper under a heated iron to avoid marking or burning the taped surface. Remove the overhanging excess tape and arris using glass-paper and a hand-sanding block.

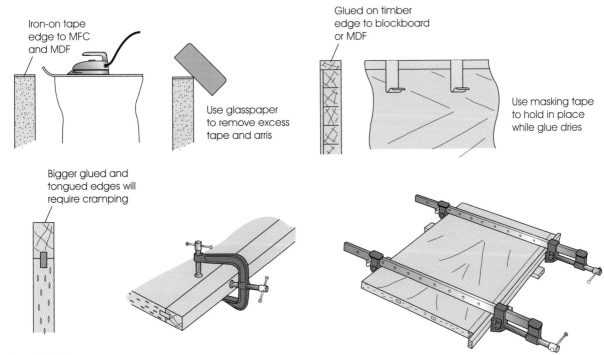

Iron-on tape edge to MFC and MDF

Use glasspaper to remove excess tape and arris

Glued on timber edge to blockboard or MDF

Use masking tape to hold in place while glue dries

Bigger glued and tongued edges will require cramping

Figure 3.57 *Applying edging*

Solid timber edging

This is more substantial and can be cut to a profile. They can be butt jointed, or use a tongue and groove for location and additional glue line for strength. Deeper edgings can be used for stiffening boards in shelves and worktops. Corner edgings can be used to join boards for carcass construction at the same time as masking the core (Figure 3.57).

- Thin solid timber edges can be glued and temporarily held in position with masking tape while the glue dries.
- Bigger lippings will require cramping by using either edging cramps or sash cramps. A batten placed under the cramp heads will spread the cramping force over the full length of the edge, as well as prevent damage.
- Use a block plane to trim glued-on timber edges flush with the board, taking care not to plane the board's surface.
- Finish off the edge by hand sanding.

Applying plastic laminate

When applying a plastic laminate finish to a surface, a balancer laminate is normally required on the reverse face of the core material to prevent distortion (Figure 3.58). Very thick cores may be used without a balancer if they are securely fixed to a sub-frame or carcass. However, the reverse face should still be sealed by a coat of adhesive to prevent moisture absorption. The procedure for applying plastic laminate will depend on the edge finish and whether a balancer is to be used. The following can be used as a general guide for a worktop with a balancer, decorative laminate top and a matching laminate front edge.

did you know?

A balancer laminate that is applied to the reverse face to prevent distortion is also termed as a compensator.

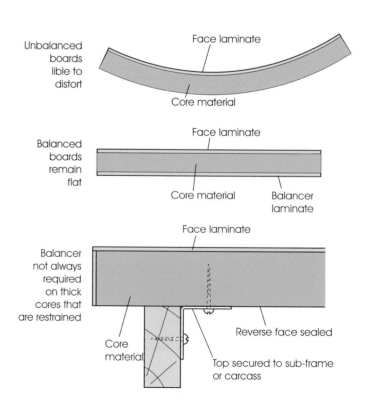

Figure 3.58 *Use of a balancer laminate*

Cutting the laminate

Always take care to avoid running your hands and fingers along the edges of laminates as they are very sharp.

The top and balancer should be cut slightly bigger than the worktop core board. Whenever possible it is best to cut laminate with the lined underside of the sheet following the longest dimension (Figure 3.59).

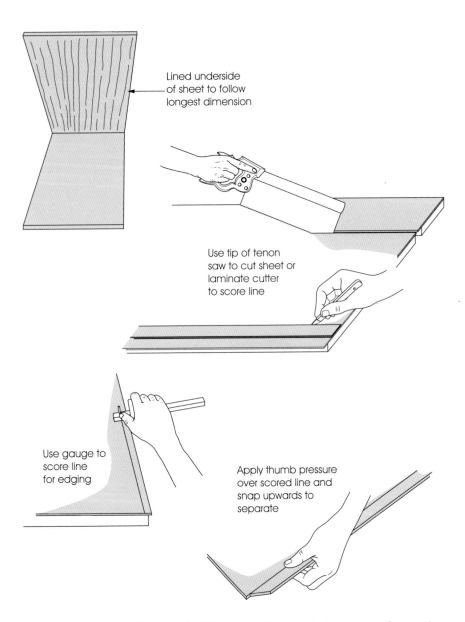

Lined underside of sheet to follow longest dimension

Use tip of tenon saw to cut sheet or laminate cutter to score line

Use gauge to score line for edging

Apply thumb pressure over scored line and snap upwards to separate

Figure 3.59 *Cutting laminate*

Support the entire sheet and either use a fine tooth tenon saw for cutting from the decorative face, or use a laminate cutter to score a line on the decorative face and break the sheet along the scored line by lifting the waste piece upwards.

Edging strips are best cut using a marking or cutting gauge:

- Set the gauge about 5 mm wider than the core board.
- Gauge along the edge of the sheet to score a line.

safety tip

Always take extra care when working with laminates as the edges are very sharp.

safety tip

Always follow the adhesive manufacturer's instructions as many adhesives give off toxic and potentially explosive vapours.

◆ Separate the strip by applying thumb pressure on the scored line, starting at one end while at the same time lifting up the edge of the strip.

◆ Carefully continue to work along the line to separate the entire strip.

Applying the balancer

◆ Dust off the balancer and the core.

◆ Lay the balancer, face-side down on the core.

◆ Apply contact adhesive to the balancer; lay off in one direction using a serrated spreader (Figure 3.60).

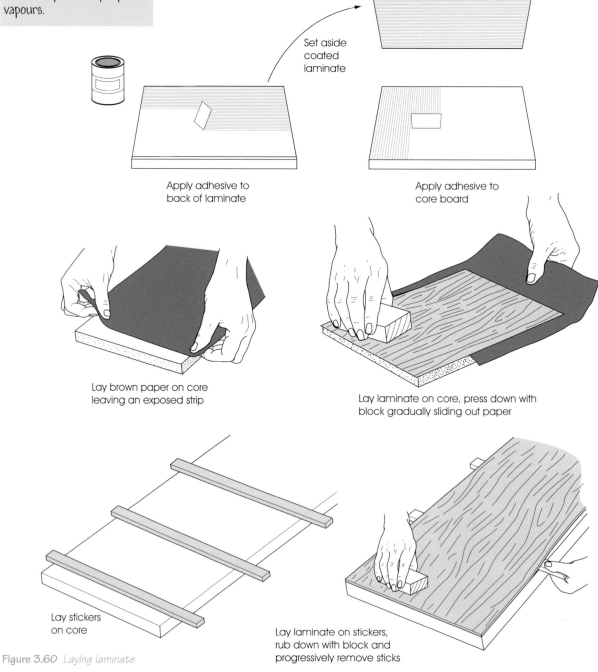

Set aside coated laminate

Apply adhesive to back of laminate

Apply adhesive to core board

Lay brown paper on core leaving an exposed strip

Lay laminate on core, press down with block gradually sliding out paper

Lay stickers on core

Lay laminate on stickers, rub down with block and progressively remove sticks

Figure 3.60 *Laying laminate*

◆ Set aside the coated balancer and apply contact adhesive to the core board, using the same method as before except this time lay off at right angles to that of the balancer. This results in a better bond than would be achieved if both are laid off in the same direction.

◆ After the manufacturer's recommended period of time (usually when the surface is touch dry) either, lay a sheet of brown paper over the entire core, leaving just a 50 mm band at one end exposed, or lay small, prepared strips of timber (stickers) across the core at about 150 mm intervals. The purpose of this is to keep the two surfaces separated until they are correctly positioned and gradually rubbed down.

◆ Position the balancer on the brown paper or stickers. When correctly aligned either:

  – press the balancer against the exposed strip, gradually slide the paper out pressing the balancer down with a block of wood to the core, working from the centre to the edges of the sheet each time, to avoid air traps, or

  – starting from one end remove the first sticker and start rubbing down as above, progressively removing the stickers as you progress.

◆ Finally, starting at the centre and working outwards, rub over the whole surface to ensure good adhesion, using a block of wood or a rubber-faced roller. Pay particular attention to the edges.

◆ Trim the edges, preferably using a guided laminate trimming cutter in a portable powered router (Figure 3.61). Where a router is not available, use a low angled block plane, file or cabinet scraper.

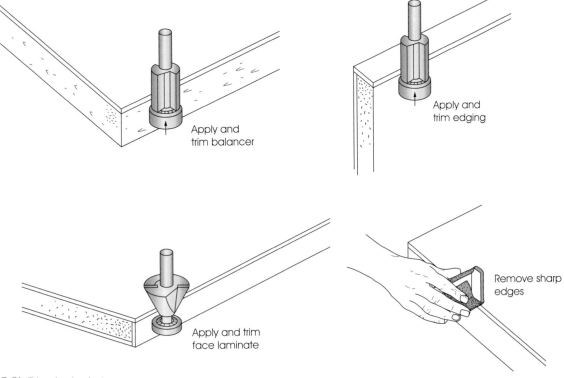

Apply and trim balancer

Apply and trim edging

Apply and trim face laminate

Remove sharp edges

Figure 3.61 *Trimming laminate*

Basic Woodworking Joints

Chapter 3

Apply edging strip

- ◆ Apply a coat of contact adhesive to the edge of the core and allow to dry. This acts as a primer to seal the absorbent surface.
- ◆ Apply contact adhesive to the back of the edging strip and a second coat to the edge of the core.
- ◆ When the adhesive is touch dry position the edging, working from one end, and keeping an even overhang on both the balancer and core face. To ensure good contact use a block of wood and tap it with a hammer along the entire edge.
- ◆ Trim the edging flush with the core face and balancer, again using a router if available.

Apply *decorative face laminate*

This stage is carried out in the same way as the balancer, except that extra care is required when hand trimming and cleaning up, so as not to spoil the decorative face. Finally, the sharp arris should be removed using a fine abrasive paper and a hand block.

activity

You have been asked to assist in the training of some school students on a work experience scheme. Compile a list of operations required to prepare some sawn timber and form a haunched mortise and tenon joint. Illustrate with sketches where appropriate and define any technical terms used.

measuring up

1. Produce a sketch to show the difference between stubbed, through, haunched and twin tenons.

2. State two reasons for using a haunch in a mortise and tenon joint.

3. When forming housing joints the depth of the housing should not exceed which of the following?
   - **a)** half of the timber thickness
   - **b)** one third of the timber thickness
   - **c)** one quarter of the timber thickness
   - **d)** two thirds of the timber thickness

4. Produce a sketch to show a scarf joint and state a typical use.

5. State the reason for applying two coats of adhesive to the edges of a chipboard worktop that is to be covered with a plastic laminate.

6. Name the joint illustrated and state a typical use.

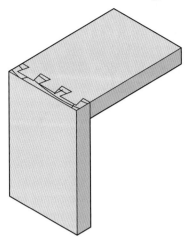

7. Produce a sketch to show two types of 'knock down' fixing which are suitable for jointing board materials at right angles.

8. Which of the following is the most suitable chisel width for mortising a 38 mm × 72 mm framing member?
   a)　6
   b)　9
   c)　12
   d)　18

9. A 'rubbed joint' has been specified for a particular job. Explain what this means.

10. Produce a sketch to show splines used to reinforce the mitre joints of a mirror frame.

4 Portable Powered Hand Tools

This chapter is intended to provide the reader with an overview of the types, uses and care of portable powered hand tools. Although its content is not assessed directly, knowledge of its contents is assumed and assessed in the following Wood Occupations units VR 05, VR 06, VR 07 and VR 08 at Level 1. Knowledge is also assessed in both Site Carpentry and Bench Joinery at Level 2.

In this chapter you will cover the following range of topics:

- Labelling on power tools
- Power supplies
- Safety procedures
- Power drills and drivers
- Portable power saws and jointers
- Planers and routers
- Sanders
- Automatic pin drivers

What's required in VR 05, VR 06, VR 07 & VR 08?

To successfully complete these units you will be required to demonstrate your skill and knowledge of the following processes:

- Setting up and using portable powered hand tools.

You will be required practically to:

- Use a range of portable powered hand tools for
  - ripping, crosscutting and forming curves;
  - mitre and compound bevels;
  - planing, rebating, grooving and moulding;
  - drilling and boring;
  - inserting screws and finishing.
- Observe safe working practices and follow instructions when maintaining, setting up, using and changing tooling for portable powered hand tools.

The woodworker has at their disposal a wide range of powered hand tools, enabling many operations to be carried out with increased speed, efficiency and accuracy.

Labelling on power tools

Power tools will have a label attached to the body (Figure 4.1) which gives the following typical details:

- Maker's name.
- The CE mark showing that the tool has been designed and manufactured to European standards.
- The double insulation mark showing that the motor and other live parts are isolated from any section of the tool that the operator can touch (for mains powered tools).
- Model and serial number, which must be quoted, when purchasing replacement parts.
- Electrical information (for mains powered tools):
    - V = voltage (electrical pressure);
    - Hz = Hertz (the frequency for AC motors);
    - A = amperes or amps (typical current flow);
    - W = watts (amount of electrical power being used).
- Electrical information (for battery power tools):
    - V = volts (electrical pressure);
    - Ah = amp-hours (capacity of battery).
- Air supply usage information (for air powered tools):
    - L/min = litres per minute (average air consumption);
    - bar = operating air pressure (1 bar is equal to atmospheric pressure).
- Other information:
    - RPM = revolutions per minute of the motor.
    - Minimum and maximum chuck opening size or depth of cut, etc., depending on the type of tool.

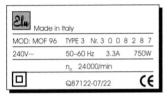

Double insulation symbol

Figure 4.1 *Typical power tool label*

Power supplies

Electricity is the main source of power supply commonly used in workshops, building sites and on customer's premises. Compressed air is used to a lesser extent. See Figure 4.2 for typical sources of power.

Electricity – is supplied from:

- the mains at 240 volts or stepped down to 110 volts via a transformer;
- a portable, petrol or diesel powered generator at 240 volts or 110 volts;
- rechargeable batteries at various voltages for portable power tools.

Compressed air – is supplied from a portable or fixed air compressor.

Electricity supplies

240 volt equipment

This is normally supplied via a 13-A, 3-pin plug and a 2 or 3-core flexible cable, which can be plugged directly into a standard mains outlet

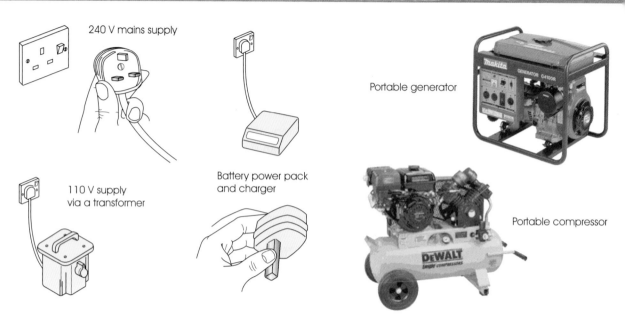

240 V mains supply

110 V supply
via a transformer

Battery power pack
and charger

Portable generator

Portable compressor

Figure 4.2 *Power supplies*

safety tip

The use of 110 volt power
supplies is safer in the event
of a fault occurring.

did you know?

Reduced voltage equipment
operating from 110 V is the
specified voltage for using
power tools on building sites.

(Figure 4.3a). The live wire (brown sheathing) is protected by a fuse, which should be matched to the ampere rating of the tool. If a fault occurs the earth wire (green and yellow banded sheathing) carries the current safely to earth. In the event of a fault, an increased flow of electricity passes through the fuse, causing the fuse to 'blow' (burn out) and cut off the electricity supply to the tool.

Double insulation – All new power tools are double insulated and may be supplied without an earth wire (Figure 4.3b). However, the live wire must still be fused to suit the ampere rating of the tool. Double insulted tools are safer than the older single insulated ones. However, both present a safety hazard to the operator if the cable carrying the electricity is damaged during the work or is faulty.

110 volt equipment

This operates via a step-down transformer with a centrally tapped earth, so that in the event of a fault, the maximum shock an operator should receive would be 55 V (Figure 4.3c). This voltage reduction provides safer operation by lowering the fault current if and when it passes through the body:

◆ Low levels of current may only cause an unpleasant tingling sensation, but the consequent momentary lack of concentration may be sufficient to cause involuntary reactions or injury (unintentional contact with moving parts, falling, etc.).
◆ Medium levels of current passing through the body, result in muscular tension and burning, in addition to the involuntary reactions or loss of concentration.
◆ High levels of current passing through the body again causes muscular tension or spasms and burning, but in addition affect the heart and can result in death or serious injury.

Since lower voltage equals lower currents equals lower risk, reduced voltage schemes are the accepted procedure for safe working.

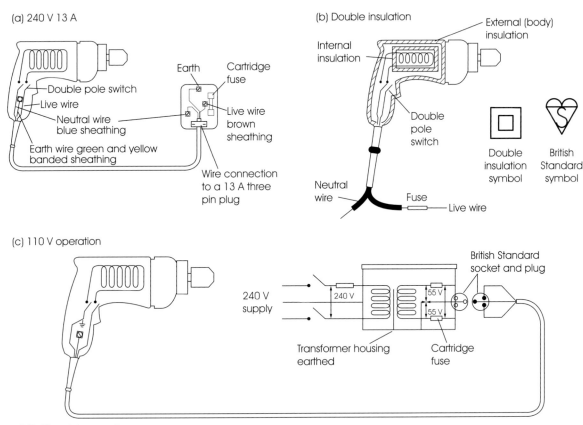

Figure 4.3 *Electricity supplies*

safety tip

RCDs are designed to trip the power supply in the event of a fault occurring.

Residual current device (RCDs)

The use of an RCD provides increased protection to the operator from electric shock. RCDs work by monitoring the flow of electricity in the live and neutral wires of the power tool circuit. In the event of a fault, an imbalance in the flow occurs causing the device to trip, cutting off the electrical supply almost instantaneously.

RCDs are available as either (Figure 4.4):

◆ a permanently wired 13-A socket outlet, with RCD fitted to a ring main circuit;
◆ an RCD adapter fitted between the tool plug and mains socket outlet;
◆ a combined RCD and plug permanently fitted to the tool power lead.

The combined RCD and plug is recommended as it is impossible to inadvertently leave it out of the system, as could be the case with the other two.

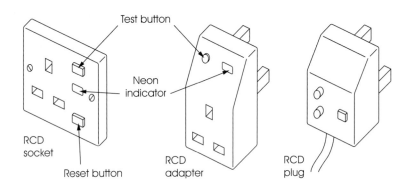

Figure 4.4 *Residual current device (RCD)*

Portable Powered Hand Tools **Chapter 4**

Before each use the 'test' button should be pressed to check the device's mechanical and electrical components. If all is working the device will 'trip', requiring the pressing of the 'reset' to restore the power supply.

Electricity supply options

Industrial power supplies

Building sites and some works may be wired using colour-coded weather-proof industrial-type shielded plugs and sockets (Figure 4.5a):

◆ Blue for 240 V and yellow for 110 V.

The position of the key and pin layout are different, making it impossible to plug into the wrong one.

When 110 V power tools are used off-site, a step-down transformer should be used to operate from the 240 V mains (Figure 4.5b). In cases where an extension lead is required this should be used on the 110 V side between the transformer and the tool. If the extension lead is stored on a drum, it must be fully unwound before use, in order to prevent overheating in the drum coil.

Portable electricity generators

Where mains electricity is not available, power for equipment can be provided by a petrol, LPG or diesel portable generator. These normally have outlets for both 240 and 110 V. Care must be taken when placing generators, in order to minimise any nuisance from emission of fumes and noise.

Battery-operated equipment

These tools are termed cordless. They are powered by removable rechargeable batteries, and are located in the tool (often the handle). When the battery has become discharged through use, it may be replaced by a recently charged spare battery.

Portable Powered Hand Tools

Chapter 4

(a) Industrial type shielded plug and socket

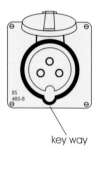

key way

key

(b) Arrangement for using 110 V power tools from 240 V mains supply

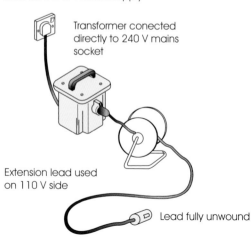

Transformer conected directly to 240 V mains socket

Extension lead used on 110 V side

Lead fully unwound

Figure 4.5 *Options for connecting to the electricity supply*

To recharge a battery – remove it from the tool and place it in a compatible battery charger. Chargers are available to run off the mains 240-V supply, from a 12-V car cigarette lighter socket or from a 110-V supply.

The time taken to recharge and the charging cycle varies between manufacturers. A typical charging cycle is:

- Fast charge – 20 minutes to 1 hour to provide the maximum charge.
- Equalisation charge – 1 to 3 hour additional charge, to further charge cells that have less charge at the end of a fast charge cycle.
- Maintenance charge – a trickle charge to maintain a full battery at peak efficiency, topping up any self-discharge.

To ensure maximum battery life, a complete three-stage charge is recommended after 20 fast charges (as above).

The amount of work a fully charged battery can do, before it requires recharging, is dependent on the type and size of the battery, the power required to drive the tool and the nature of the work being undertaken.

Take care when handling and charging batteries to retain maximum life:

- Fully charge battery before storage, but note that they will still start to self-discharge immediately.
- Store battery in a cool, dry environment, between 4 and 40°C is best, so as not to let the battery freeze.
- Remove the battery during transportation, as the accidental continuous switch operation (when in a tool bag) can cause permanent damage to the battery.
- Allow a rapidly discharged battery to 'cool down' before recharging.
- Recharge the battery when the tool no longer performs its intended use.

did you know?

The frequent overloading or stalling of a tool will reduce the life of the battery.

Compressed air (pneumatic) equipment

Air above the atmospheric pressure of 1 bar is used as the power source for pneumatic tools. Air compressed to between 5 and 8 bar is used either to turn vanes fixed to a drive shaft to give rotary motion for drills, etc. or a piston to give the motive force for driving nails and staples. Compressors used to supply the compressed air may be a permanent fixture in a workshop or a portable unit for site-work and work at customer's premises.

Air compressors consist of the following:

- Petrol, LPG, diesel or electrically powered motor.
- Air pump, where the air is compressed by a piston and passed into a receiver (Figure 4.6). The body of the pump is normally finned and should be well ventilated as heat is generated during compression.
- Air receiver in the form of a storage tank for the compressed air, which evens out the pulsating pump delivery so that the air is available for use at a constant pressure. A drain-off point is included in the bottom of the tank to periodically drain off condensation.

Delivery of compressed air to a power tool is via the following:

- A system of ridged airlines to transport the compressed air to the fixed outlet.

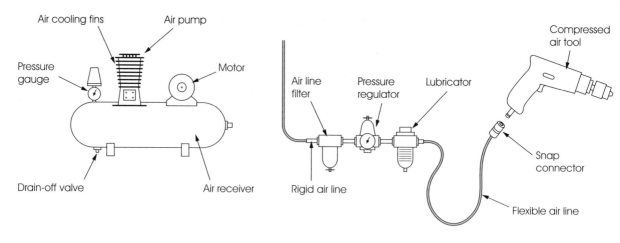

Air cooling fins   Air pump

Pressure gauge   Motor

Drain-off valve   Air receiver

Air compressor

Compressed air tool

Air line filter   Pressure regulator   Lubricator

Rigid air line

Snap connector

Flexible air line

Compressed air supply system

Figure 4.6 Compressed air equipment

◆ A valve and connector for a flexible hose to the tool. The size of the line and hose must be compatible with the tool used. Any variation in the bore of the line will cause a variation in the power supplied to the tool.

Tools used from portable compressors will be supplied directly from the compressor via a flexible hose. Ideally the compressor should be positioned close to the work, as long hoses cause pressure drop through internal friction and also create potential tripping hazards.

For maximum efficiency the air supply system should incorporate, as close to the tool as possible:

◆ a filter to remove any moisture from the air;
◆ an adjustable pressure regulator to control the air pressure reaching an individual tool;
◆ a lubricator to allow a controlled amount of lubricant to reach the tool being used.

Some tools do not require air lubrication and this should be turned off, as it may perish their internal seals, causing failure. Check manufacturer's instructions when in doubt.

Safety procedures

Compressed air equipment

Compressed air equipment can, if correctly used, be perfectly safe. However, if misused, they can cause severe personal injury. Compressed air entering the body causes painful swelling. If it is allowed to enter the bloodstream it can make its way to the brain, burst the blood vessels and cause death.

The following safety points must be observed whenever compressed air equipment is used:

◆ The compressor must be in the control of a fully trained competent person at all times.

safety tip

Always follow the safety procedures when using portable power tools.

Chapter 4 Portable Powered Hand Tools

Ear muffs

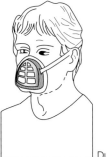

Dust mask

Respirator

Goggles

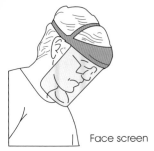

Face screen

Figure 4.7 *Personal protective equipment (PPE)*

◆ All equipment must be regularly inspected and maintained.
◆ Training must be given to all persons who will use compressed air equipment.
◆ Position portable compressors in a well-ventilated area.
◆ Ensure all hose connections are properly clamped.
◆ Route all hose connections to prevent snaking, risk of persons tripping or traffic crossing them, as any squeezing of the hose causes excess pressure on the couplings.
◆ Always isolate the tool from the air supply before investigating any fault.
◆ Never disconnect any air hose unless protected by a valve.
◆ Never use an air hose to clean away waste material, or anything else that may result in flying particles.
◆ Never use compressed air to clean down yourself or anyone else, as this carries a great risk of injury to the eyes, ears, nostrils and rectum.

General power tool safety

◆ Never use a power tool unless you have been properly trained in its use by a competent person.
◆ Never use a power tool unless you have your supervisor's permission.
◆ Always select the correct tool for the work in hand (if in doubt consult the manufacturer's instructions).
◆ Ensure that an electric power tool and supply are compatible.
◆ Ensure that the cable or air hose is:
  – free from knots and damage;
  – firmly secured at all connections;
  – unable to come into contact with the cutting edge or become fouled during the tool's operation.
◆ Route cables and hoses with care.
◆ Before making any adjustments, always disconnect the tool from its power supply.
◆ Always use the tool's safety guards correctly and never remove or tie them back.
◆ Never put a tool down until all the rotating parts have stopped moving.
◆ Always wear the correct personal protective equipment (PPE) for the job (Figure 4.7). These may include:
  – ear muffs;
  – safety goggles or face screen;
  – dust mask or respirator.
◆ Always use dust extraction equipment when operating power tools that produce dust, chips and other waste (Figure 4.8). This is in addition to and not in place of PPE.
◆ Loose clothing and long hair should be tied up so that they cannot be caught up in the tool. In the event of entanglement, switch off the power supply immediately.
◆ All power tools should be properly maintained and serviced at regularly intervals by a suitably trained person. Never attempt to service or repair a power tool yourself. If it is not working correctly or its safety is suspect, return it to the storeman with a note stating why it has been returned.

Portable Powered Hand Tools

Chapter 4

Figure 4.8 *Industrial vacuum cleaner/extractor*

did you know?

All power tools should be returned to the stores for inspection at least once every seven days.

◆ Ensure that the material or workpiece is firmly cramped or fixed in position so that it will not move during the tool's operation.
◆ In general, compressed air tools must be started and stopped under load, whereas electric tools must not.
◆ Never use an electric tool where combustible liquids or gases are present.
◆ Never carry, drag or suspend a tool by its cable or hose.
◆ Think before and during use. Tools cannot be careless, but their operators can. Most accidents are caused by simple carelessness.

Hazards associated with vibration

Exposure to vibration can cause permanent damage to a power tool operator. Effects include impaired blood circulation, damage to nerves and muscles and damage to bones in the hands and arms.

Hand-arm vibration syndrome (HAVS)

This is likely in any process where hands are exposed to vibrations from vibrating tools or workpieces. The effect of the vibration dose received by an operator over a day depends on the:

◆ vibration frequency;
◆ duration of exposure;
◆ exposure pattern;
◆ grip and force required guiding the tool or workpiece.

Precautions should be taken where:

◆ any tingling or numbness is felt after 5–10 minutes of use;
◆ high-risk operations are carried out, such as the use of hand-held sanders, hand-fed or hand-held circular saws, pneumatic nailing or stapling tools.

Employers should take preventive measures to reduce vibration, such as limiting the duration of exposure, the selection of suitable work methods, use of reduced vibration tools and appropriate PPE. In addition, health surveillance of people who are exposed to vibration should also be undertaken, especially where high levels or long durations are involved.

Manufacturers have a duty to reduce vibration levels and supply vibration data for their equipment. Figure 4.9 illustrates a typical labelling system for use particularly on power tools, which give the operator guidance on vibration levels and recommended daily maximum duration of use.

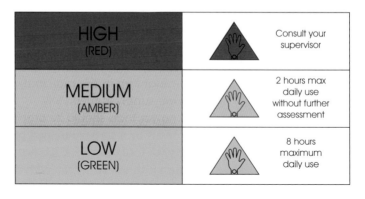

HIGH (RED)		Consult your supervisor
MEDIUM (AMBER)		2 hours max daily use without further assessment
LOW (GREEN)		8 hours maximum daily use

Figure 4.9 *Manufacturer's labelling system to warn users of vibration levels*

Power drills and drivers

As with all equipment, electric power tool manufacturers supply tools for various markets, ranging from the cheap, almost 'throw away' type for the bottom end of the 'do it yourself market' suitable only for light/occasional use, through to the more sophisticated and powerful industrial models, which will stand up to constant heavy use.

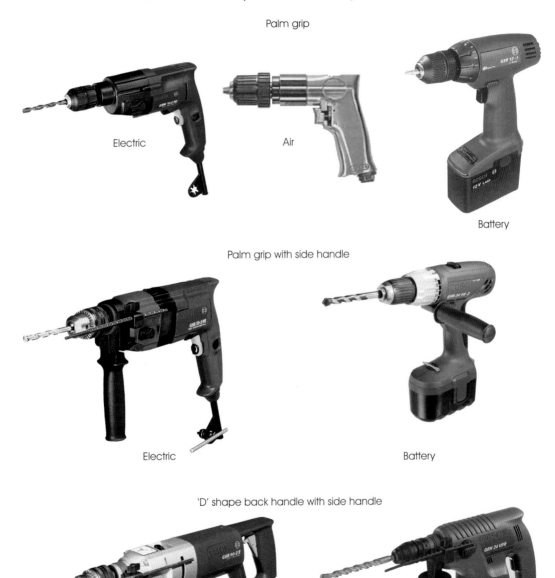

Palm grip

Electric

Air

Battery

Palm grip with side handle

Electric

Battery

'D' shape back handle with side handle

Electric

Battery

Figure 4.10 *Types of power drill*

The power drill is the most common type of portable power tools in use. Your choice of drill (Figure 4.10) will depend on the following factors:

Portable Powered Hand Tools **Chapter 4**

◆ power supply available;
◆ amount of use;
◆ size of holes required;
◆ material being drilled.

Types of power drill

Palm grip handle – used for lighter drills, and for single-handed use. The design ensures that the pressure is exerted directly in line with the drill or screwdriver bit. On heavy drills a secondary side handle is fitted near the chuck, for two-handed operation.

'D' shape back handle – used for large, heavy drills, along with an adjustable front handle for two-handed operation. This design is suited to percussion or hammer action drills and for boring in masonry and concrete, with special percussion drill bits.

Chucks

Drill and screwdriver bits are held securely in the chuck, normally by three self-centring jaws that grip the shank of the bit. Traditionally, chucks are operated with a toothed key to open and close the jaws (Figure 4.11). Other more recent chucks are 'keyless' being operated by gripping and turning the cylinder collars. Some of the larger drills are fitted with a fast action chuck (SDS). When the chuck collar is pulled back it automatically opens, enabling a special bit with grooved shanks to be inserted.

Chuck capacity – describes the range of drill bit sizes that the chuck can accept. This corresponds to the maximum size of hole that the tool is capable of boring in steel. Small drills have a chuck capacity of 1 to 10 mm in diameter and larger ones 1.5 to 13 mm. Both sizes may be used to bore larger holes in wood (mainly) when fitted with reduced shank drills or spade bits, etc.

'Keyless' three jaw chuck

Key operated three jaw chuck

Chuck key

Figure 4.11 *Drill chucks*

Drill features

Speed selection – Drills perform better when the speed of the bit can be adjusted to suit the size of hole and material being worked (Figure 4.12). In general use a fast speed for boring holes in wood and a slower speed for drilling holes in metal, masonry and for driving woodscrews.

◆ Basic drills may only have one non-load speed (2400 rpm). Other basic drills may be equipped with gearing enabling two fixed non-load speeds (900 and 2400 rpm) via a lever or switch.
◆ Variable speed drills vary from zero to maximum (0 to 2400 rpm) according to the pressure applied to the trigger. Some are equipped with a dial to pre-set optimum speeds, by limiting the trigger movement. Variable speed drills may also be equipped with two mechanical gears; the slower speed provides more torque (rotating power).
◆ A 'soft start' is a useful drill feature as it minimises the initial jolt of a high speed electric motor, which can cause drill-bit skidding and damaged screw heads.

Reverse action – Many air and mains electric power drills and all battery drills are fitted with a switch, to change the direction of rotation for removing screws.

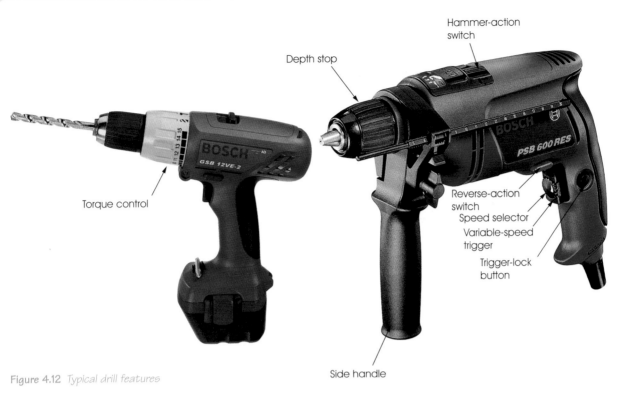

Hammer-action switch

Depth stop

Torque control

Reverse-action switch

Speed selector

Variable-speed trigger

Trigger-lock button

Side handle

Figure 4.12 *Typical drill features*

Torque control – Most battery drills are equipped with adjustable torque control enabling screws of different sizes to be driven to flush or below the fixing surface. Adjustment is via a collar behind the chuck. Use the lower torque numbers for smaller screws and the higher numbers for large screws or when driving screw heads below the surface.

Hammer action

Both mains electric and battery-powered drills may be equipped with a hammer action, enabling easy boring into brick, stone and concrete. Operated by switch before or during operation, the hammer action puts typically 4000 blows per minute (BPM) behind the rotating drill bit to break up the masonry. It is essential that special percussion drill bits are used for this work.

Side handle and depth stop

did you know?

When boring holes, ensure the torque control is set to the drill position to avoid slipping.

Most mains electric drills and some battery drills can be fitted with an additional side handle for two-handed control. Integral depth stops can be adjusted to come into contact with the workpiece when the drill has reached the required depth.

Trigger lock

A button adjacent to the trigger can be pushed in with the drill running, to lock the trigger for continuous drilling. Re-squeezing the trigger releases the lock and stops rotation.

Angle drilling

Purpose-made angle drills or attachments for standard drills enable holes to be bored in confined spaces (Figure 4.13).

Angle drill for confined spaces

Angle drilling attachment

Figure 4.13 *Angle drill and angle attachment*

Drill stand

For repetitive drilling work and light mortising a drill can be fitted into a drill stand (Figure 4.14). In order to comply with safety requirements the stand is fitted with a retractable chuck and drill guard which must always be in position when drilling. Various size chisels along with their auger bits are available for use in a drill stand for mortising.

Twist drills, bits and accessories

Twist drills

Twist drills and bits for use in power drills (Figure 4.15) should be high quality to withstand the heat generated when drilling at speed. They are available in sets ranging in size from 1 mm, increasing by 0.5 mm, to 13 mm in diameter. Carbon steel drills are suitable for woodwork. High-speed steel drills (HSS), which stay sharper longer, are considered better quality and are suitable for drilling wood, metal and plastics. Titanium-coated HSS drills are also available for extended working life. Larger twist drills from 13 to 25 mm are made with reduced shanks to fit standard drill chucks. Standard length twist drills, termed 'jobbers', range from 35 mm in length for the 1 mm diameter up to 150 mm in length for the 13 mm diameter. Extra length twist drills are available for drilling deeper holes.

Twist drills are not easily centred, so it is worthwhile centre-punching the centre of the hole first to avoid skidding off the mark.

Brad point bits

Also known as lip and spur or dowel bits. These are wood bits, and have a centre point to prevent skidding off the mark when the power is applied. These are available in a range of sizes from 3 mm to 20 mm in diameter. Sizes over 13 mm have reduced diameter shanks enabling the use of a standard chuck.

Spade bits

Also known as flat bits, these are relatively cheap bits for drilling larger diameter holes in wood. They are available from 6 mm to 50 mm in diameter. Make sure the point is fully inserted in the wood before starting to drill.

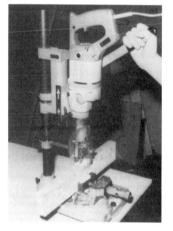

Figure 4.14 *Drill stand*

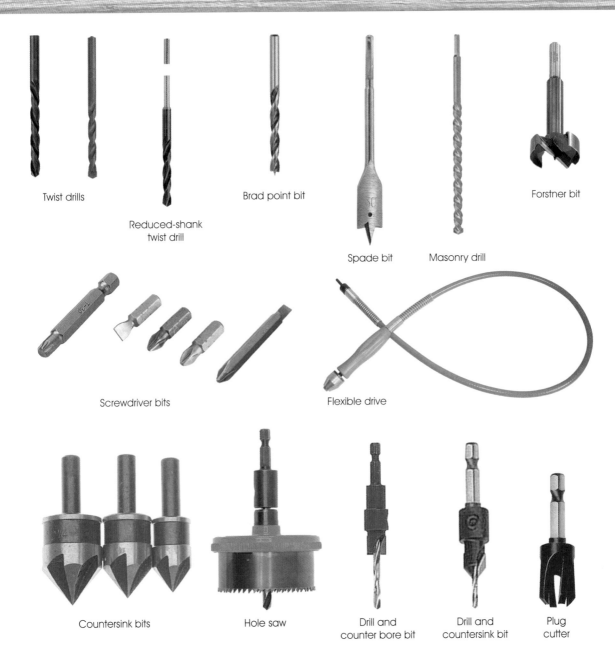

Twist drills

Reduced-shank
twist drill

Brad point bit

Spade bit

Masonry drill

Forstner bit

Screwdriver bits

Flexible drive

Countersink bits

Hole saw

Drill and
counter bore bit

Drill and
countersink bit

Plug
cutter

Portable Powered Hand Tools

Chapter 4

Figure 4.15 *Twist drills, bits and accessories*

Masonry drills

These are similar to twist drills but have a tip of tungsten carbide fitted into a slot on the cutting end used for boring into masonry.

Forstner bits

Superior quality bits which leave a clean, accurate and if required flat bottom or blind hole. They are available in sizes from 10 mm to 50 mm in diameter. Unlike other drills which are guided by their centre point or spiral, Forstner bits are guided by their rim, enabling them to bore holes that run over the edge of a workpiece and to bore overlapping holes.

Countersink bits

A countersink bit makes a countersunk or tapered recess to accommodate the head of a woodscrew. A pilot and clearance hole should be drilled first

safety tip

When used in a hammer action/precision drill ensure the tipped drill is recommended by the manufacturer to withstand the vibration produced, otherwise the tip is likely to shatter.

with a twist drill. The countersunk bit is located in the clearance hole and run at high speed for a clean finish.

Combined drill and countersink bits – are available in a range of sizes to suit the most common woodscrews. They drill a pilot hole, shank clearance hole and countersink in one operation. Versions are available to drill a counter-bore hole to accept a wooden plug which conceals the screw head.

Counter bore bits – drill a pilot, clearance and counter-bore hole. The counter-bore hole can be subsequently filled with a wooden pellet to conceal the screw head.

Plug cutters – should match the counter-bore bit size. They cut a wooden plug from timber that closely matches the grain and colour of the work. Plug cutters can only be used effectively in a pillar drill or drill stand attachment.

Concealed hinge boring bits

These are available in 25, 30 and 35mm diameters, for boring concealed hinges. Similar to a Forstner bit, many have a tungsten carbide tipped (TCT) cutting edge. For control it is best used in a pillar drill or with a drill stand.

Hole saws

These consist of a cylindrical shaped saw blade, which is held in a backing plate, secured to a twist drill, passing through its centre. Hole saws are typically available from 14mm to 150mm in diameter. As the saw perimeter moves faster than the drill perimeter, select a slower speed and steadily feed the saw into the work.

Screwdriver bits

A range of slot, cross and security head tips are available as screwdriver bits for power drills and drivers. When driving or removing screws, the slowest speed should be selected and pressure maintained during the entire operation to avoid the bit jumping out of the screw.

Flexible drive

This allows use of a drill in awkward places. It consists of a 1m drive cable sheathed in a flexible casing, with a 6mm capacity chuck at one end of a spindle and at the other end a spindle for securing into a standard power drill chuck.

Hinge drill

This is a twist drill secured within a spring-loaded housing having a chamfered end to suit the hinge countersink holes. The hinge is used as the template, with the chamfered end positioned in the screw hole; pressure is applied to bore perfectly positioned pilot holes.

Portable drill stand

This telescopic device attaches to the standard 43mm drill collar and enables the precision drilling of holes at right angles; it can also be adjusted for holes up to 45° and is ideal for use with plug cutters and concealed hinge boring bits.

Collar size

Many drills have an international standard 43 mm collar directly behind the chuck. This enables the drill to be used with a range of attachments.

Operation of power drills

During use:

- Hold the drill firmly.
- Cramp the workpiece.
- Do not force the drill.
- Remove the drill from the hole frequently to clear the dust and allow it to cool.
- Drill a small pilot hole first to act as a guide when drilling larger holes.
- Reduce the pressure applied on the drill as it is about to break through to avoid snatching or twisting.

Power-operated screwdrivers

In addition to the dual use of power drills as screwdrivers there is a range of dedicated power screwdrivers available (Figure 4.16). The body and motor is similar to that of an electric power drill, although a reduction gear is fitted to give the optimum speed for screw driving. Where a two-speed tool is used, the slow speed should be used for woodscrews and the high speed used for self-tapping screws into thin metal-work.

The front housing of the tool holds the screwdriver bits and contains a clutch assembly that operates in two stages:

- The tool motor will run but the screwdriver bit will not rotate until sufficient pressure is exerted on the tool to enable the clutch to operate and engage the main drive.
- When the screws have been driven and are tight in the required position, the second stage of the clutch operates and stops the screwdriver bit rotating.

did you know?

Most tools are manufactured with a reverse gear to enable screws to be removed.

Dry wall screwdrivers – have a single clutch, which allows screws to be driven through plasterboard panels into the studwork without pre-drilling. Bugle-head screws are normally used and the clutch will disengage when the head is just below the surface but without breaking through the paper covering.

Power-screwdrivers – are equipped with adjustable clutch settings, enabling different size screws to be driven to a predetermined depth below the work surface before the clutch disengages.

Automatic feed power screwdrivers – in these, the screws are attached to a plastic cartridge belt, which automatically feeds the screws into the driving position, one after the other. Collated screws are available in a range of sizes in both woodscrew and drywall patterns. Typically each cartridge contains 1000 screws.

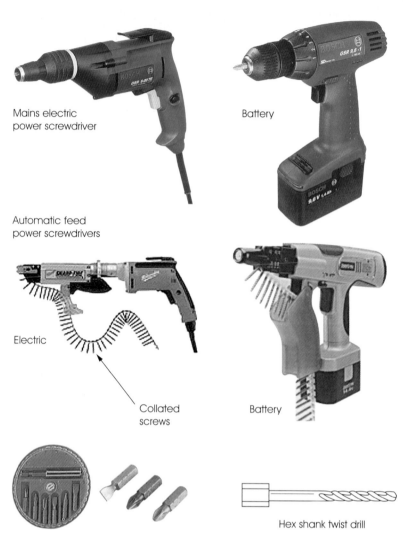

Figure 4.16 *Power-operated screwdrivers*

Screwdriver bits and sockets

Various screwdriver bits and sockets are available to suit different types and sizes of screws. It is a simple operation to change the type of bit when required. The hexagonal shank of the bit is simply pushed into the front housing of the tool and retained in position by a spring-loaded steel ball, which locates in a groove around the top of the shank. The bit is removed by simply pulling it out of the front housing. In addition hex shank twist drills are available for use as pilot drills in power screwdrivers.

Operation of power screwdrivers

- Select a screwdriver bit that is compatible with the tool and screw being used.
- Drill a pilot hole, clearance hole and countersink where required before screwing, to avoid overloading the motor and splitting the material.
- Maintain a steady, firm pressure on the screwdriver so that the bit cannot jump out and damage the screw head or the workpiece.

Portable power saws and jointers

There are many types of power saw available to the woodworker each with its own specific range of functions.

Circular saws

Often termed skill saws (Figure 4.17), these are used for a wide range of sawing operations including cross-cutting, rip sawing, bevel and compound bevel cutting, rebating, grooving trenching and the cutting of sheet material. Blade diameters range from 130 mm to 240 mm, giving a vertical maximum depth of cut ranging from 40 mm to 90 mm. Angle cuts up to 45° are possible but at reduced maximum depths. Various teeth profiles are available to suit the work in hand. However, the use of tungsten-carbide tipped saw blades is preferable in all situations, but particularly when cutting plywood, MDF, chipboard, fibreboard, plastic laminates and abrasive timbers.

Operation of a circular saw

Ensure that the blade guard has sprung back over the blade before putting the tool down.

- Select and fit the correct blade for the work in hand (rip, cross cut, combination or tungsten tipped, etc.).
- Adjust the depth of the cut so that the teeth project about half their length through the material to be cut (Figure 4.18).
- Check the blade guard is working properly. It should spring back and cover the blade when the saw is removed from the timber.
- Check the riving knife is rigid and set at the recommended distance from the blade.
- Set the saw to the required cutting angle. This is indicated by a pointer on the pivot slide.
- Insert the rip fence (if required) and set to the width required. When cutting sheet material or timber where the rip fence will not adjust to the required width, a straight batten can be temporarily fixed along the board to act as a guide for the sole plate of the saw to run against.

Figure 4.17 Circular saws

Portable Powered Hand Tools

Chapter 4

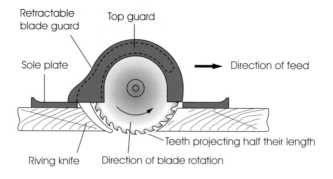

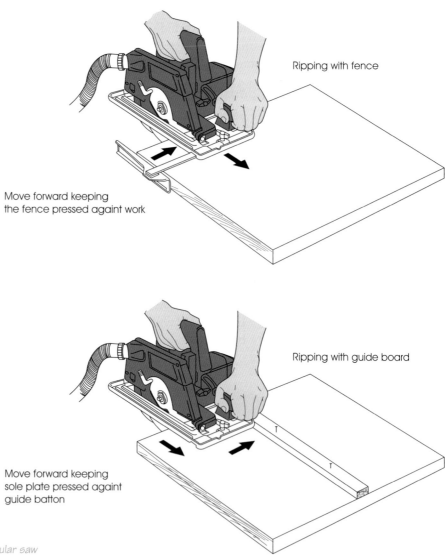

Figure 4.18 *Operation of the circular saw*

◆ Check to ensure that all adjustment levers and thumbscrews are tight.
◆ Ensure that the material to be cut is properly supported and securely fixed down. As the saw cuts from the bottom upwards, the face side of the material should be placed downwards. This ensures that any breaking out which may occur does not spoil the face of the material.
◆ Rest the front of the saw on the material to be cut. Depress the safety lock with your thumb and pull the trigger to start the saw. Circular saws should not be fitted with a trigger lock, which enables continuous running.
◆ Allow the blade to reach its full speed before starting to cut. Feed the saw into the work smoothly and without using excess pressure. The blade guard will automatically retract as the saw is fed into the work.
◆ If the saw binds in the work, ease it back until the blade runs free.
◆ When the end of the cut is reached, remove the saw from the work, allowing the blade guard to spring back into place and then release the trigger.

Cross-cutting

Square and bevelled cross-cutting can be undertaken by clamping or pinning a guide batten to the workpiece, with the blade set in the vertical position (Figure 4.19). The blade is set at an angle to the baseplate for compound bevel cutting.

Rebating – is done using the rip fence as a guide. Make two saw cuts along the grain, one from the edge and the second from the face (Figure 4.20). The blade must be set to the required depth.

Plough grooves – are again formed using the rip fence as a guide. Set the fence to cut the two sides of the groove. Reset the fence and gradually remove the waste by making successive passes with the saw.

safety tip

Do not release the trigger before the end of the cut has been reached.

Portable Powered Hand Tools **Chapter 4**

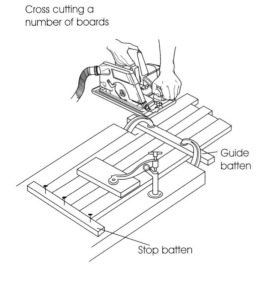

Cross cutting a number of boards

Guide batten

Stop batten

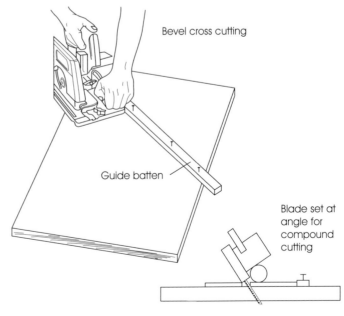

Bevel cross cutting

Guide batten

Blade set at angle for compound cutting

Figure 4.19 *Cross-cutting*

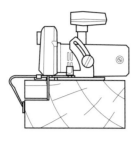

Rebate cut in two stages

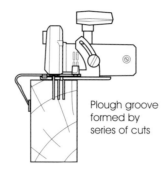

Plough groove formed by series of cuts

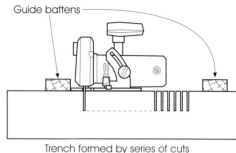

Guide battens

Trench formed by series of cuts

Figure 4.20 Rebating, grooving and trenching

Trenching – is carried out by making a series of parallel saw cuts across the grain, with the blade set to the required depth. Use a guide batten on either side of the trench and make successive passes of the saw to remove the waste.

Pocket cutting

This is a cut that starts and finishes within the length or width of a board or floor, etc. This cut that can be used when cutting traps for access to services in completed floors. When cutting pockets the following stages should be followed (Figure 4.21). This is in addition to the normal operating stages.

◆ Mark the position of the pocket to be cut; if on a floor, check that the lines to be cut are free of nails, etc.
◆ Adjust the depth of the cut so that the saw blade will penetrate the floorboards by less than 1 mm. If the blade is allowed to penetrate deeper there is a danger that the saw might cut into any services, which may be notched into the tops of the joists, e.g. electric cables, water and gas pipes.
◆ Tack a batten at the end of the cut to act as a temporary stop.
◆ Place the leading edge of the saw's sole plate on the work surface against the temporary stop.
◆ Partially retract the blade guard using the lever and start the saw.
◆ Allow the blade to obtain its full speed and gently lower the saw until its sole plate is flat on the work surface.
◆ Release the trigger, allow the blade to stopand remove the saw. Turn it around and complete the cut in the opposite direction to the corner.
◆ Repeat the previous stages on the other three sides to complete the access trap.

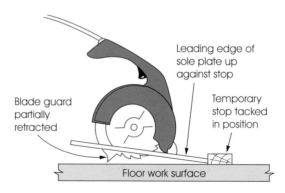

Leading edge of sole plate up against stop

Blade guard partially retracted

Temporary stop tacked in position

Floor work surface

Figure 4.21 Circular saw ready for pocket cutting

Figure 4.22 *Jig (reciprocating)*
saws

did you know?

Some models of jig saw are equipped with replaceable inserts, which fit into the sole to fill the space around the blade. This supports the material either side of the kerf, thus minimising chipping.

Jig saws

These are also known as reciprocating saws (Figure 4.22). Although they may be used with a fence for straight cutting they are particularly useful for circular, shaped and pierced work. In addition many models have adjustable sole plates, which allow bevel cutting to be carried out.

Basic jig saws have a simple up and down action with the blade cutting on the upward stroke. Other better quality saws are equipped with a pendulum action that advances the blade into the work on the upward cutting stroke and moves it away on the down stroke. This cutting action prolongs the life of the blade by minimising teeth wear and friction on the down stroke. It also helps to clear the kerf of waste.

As the blade cuts on the upward stroke, it tends to chip or splinter on both sides of the kerf on the upper surface of the workpiece. Either apply a length of masking tape over the line to be cut, or cut with the 'good' or finished face to the underside (Figure 4.23).

A range of different blades are available for cutting a wide variety of materials.

Operation of jig saw

- ◆ Select the correct blade for the work in hand.
- ◆ Select the correct speed: slow speed for curved work and metal cutting, high speed for straight cutting.
- ◆ Ensure that the material to be cut is properly supported and securely fixed down.
- ◆ Rest the front of the saw on the material to be cut and pull the trigger to start the saw.
- ◆ When the blade has reached its full speed, steadily feed the saw into the work, but do not force it as this may cause it to wander.
- ◆ When the end of the cut is reached, release the trigger, keeping the sole plate of the saw against the workpiece, but making sure the blade is not in contact.

Pocket cutting

To cut a pocket or aperture in a board, a starter hole for the blade is first bored in the waste (Figure 4.24). Insert the blade in the hole, switch on and cut along the first line into the corner, back-up the saw about 25 mm and make a curved cut on to the second line of the aperture. Repeat this operation at each corner until the aperture is complete. Finally, go back to each corner and remove the triangular waste piece by sawing back in the opposite direction.

Alternatively, a plunge cutting technique can be used to cut apertures without the need for a starter hole. After setting the saw up and following the safety precautions as before:

- ◆ Hold the saw over the line to be cut and tip forward until the front edge of the sole plate rests firmly on the workpiece.
- ◆ With the blade clear of the work surface, and keeping a firm grip on the saw, pull the trigger. Pivot the saw from the front edge so that the blade cuts into the surface and full penetration is made.

Chapter 4 Portable Powered Hand Tools

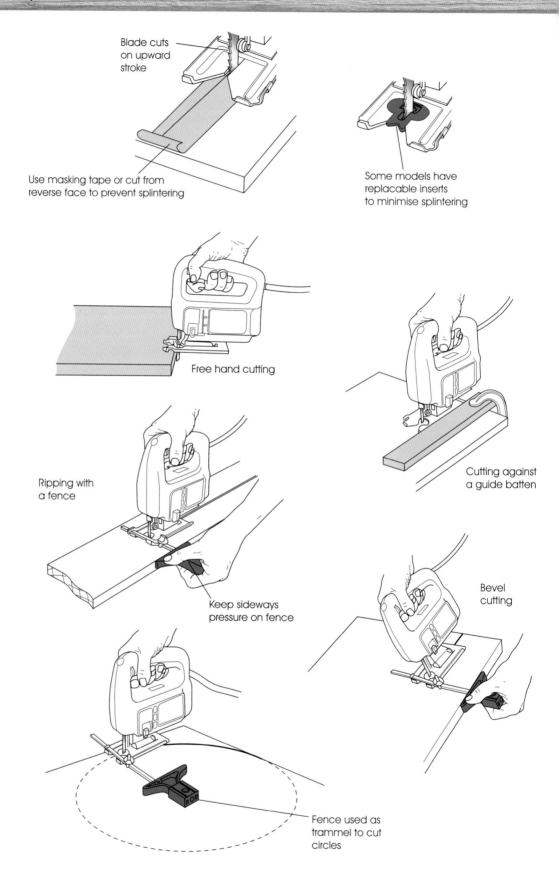

Figure 4.23 *Operation of jig saw*

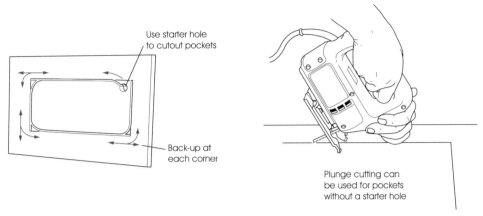

Use starter hole to cutout pockets

Back-up at each corner

Plunge cutting can be used for pockets without a starter hole

Figure 4.24 *Pocket cutting using a jig saw*

◆ Proceed with normal cutting to complete the shape. If the pocket being cut is square or has sharp corners, it will be necessary to repeat the plunge cut along its other sides.

Mitre saw

Also known as a chop saw, mitre saws consist of a circular saw that pivots from a pillar above a circular baseplate (Figure 4.25). The baseplate and saw can be rotated in either direction from the square by up to 45° in relation to the fence. This enables square and mitred cross-cuts to be made. In addition the motor and blade assembly can be adjusted so that the blade can be set at a bevel angle to the baseplate. Some models can be bevelled in both directions, others to the left only due to the motor projection. Standard adjustment is between 0° and 45° from the vertical. This adjustment allows bevelled and compound mitred cross-cuts to be made.

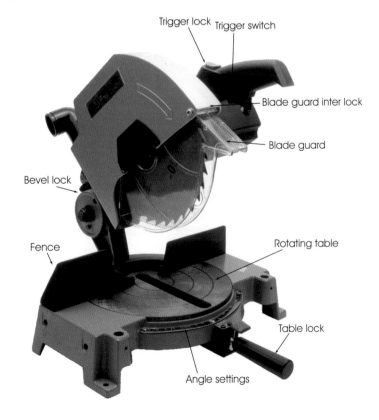

Trigger lock Trigger switch

Blade guard inter lock

Blade guard

Bevel lock

Fence

Rotating table

Table lock

Angle settings

Figure 4.25 *Mitre saw*

Portable Powered Hand Tools **Chapter 4**

The operator starts the saw using the handle trigger and blade guard interlock device. When the saw is running at full speed, a downward movement on the handle retracts the blade guard and brings the blade into contact with the workpiece to make the cut.

The size of material that can be cut is limited by the size of the saw blade. Typical cross-sectional capacities for a 190 mm blade are (Figure 4.26):

- 105 mm × 50 mm for square cross-cutting;
- 65 mm × 50 mm for 45° mitring;
- 35 mm × 35 mm for 45°/45° compound mitres.

Models with larger diameter blades will have a bigger cutting capacity. Some more advanced saws are available with a pull-out function in addition to the plunge or chop action. This allows the cross-cutting of wider boards.

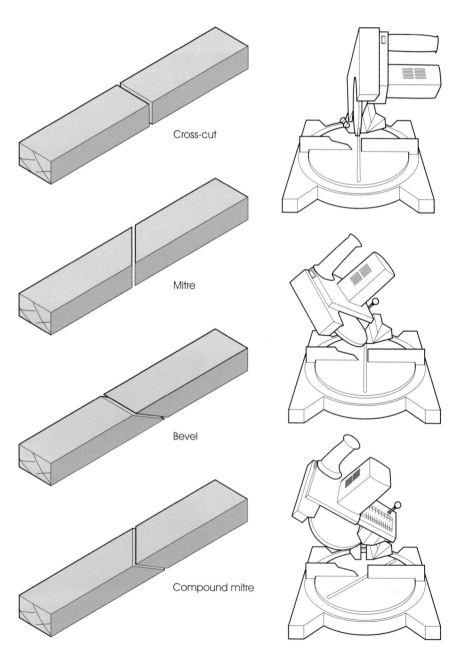

Cross-cut

Mitre

Bevel

Compound mitre

Figure 4.26 *Types of cut using a mitre saw*

Most mitre saws are equipped with dust extraction facilities, which should be used at all times. In addition you should always wear ear defenders, eye protection goggles and a dust mask when using a mitre saw.

Operation of a mitre saw

◆ Cross-cuts at 90° to the edge of the workpiece are made with the mitre table set at 0°.
◆ Mitred cross cuts are made with the table set above 0° in either direction.
◆ Bevel cuts are made with the mitre table set to 0° and the blade set at an angle between 0° and 45°.
◆ Compound mitred cross-cuts are a combination of a mitre and a bevel cut.
◆ Rotate the mitre table, so that the indicator aligns with the required angle. Lock the table in position.
◆ Check all adjustments are tight; perform a dry run of the cutting action to check for problems before plugging into the power supply.
◆ Position the workpiece on the base table with one edge against the fence. Where the timber is warped put the convex edge to the fence; putting the concave edge to the fence is dangerous, as it can result in snatching of the workpiece and jamming the blade. Long lengths will require an end support, level with the saw table.
◆ Preferably cramp the workpiece to the fence. Smaller sections may be held in position with the left hand. However, ensure it is well clear of the saw's cutting area.
◆ Grip the handle firmly, squeeze the trigger and allow the blade to reach its maximum speed.
◆ Operate the blade guard interlock lever and slowly lower the saw blade into and through the workpiece.
◆ Release the trigger and allow the saw blade to stop rotating before raising it from the work. Short off-cuts may be dangerously projected across the workshop if the blade is lifted whilst still running.
◆ Always disconnect from the power supply before making any further adjustments or changing the saw blade.

Biscuit jointer

This is a miniature plunging circular saw, developed principally to make a form of tongued and grooved joint for furniture and cabinetwork (Figure 4.27).

Figure 4.27 Biscuit jointer

The joint works like a dowel joint, except that compressed beech oval biscuits fit into matching saw slots. When glue is applied to the joint, the biscuit expands to fill the slot producing an easy fitting butt joint (Figure 4.28). Biscuit jointers can also be used to cut grooves and trim sheet material.

Biscuit jointing is ideal for both framed and cabinet carcass construction, using solid timber and board material. Joints can be butted, mitred or edge-to-edge.

Operation of a biscuit jointer

◆ Mark the centre line of the joint on both pieces (Figure 4.29).
◆ Mark the biscuit slot centres along the centre line, typically spaced about 100 mm to 150 mm apart.
◆ Set the depth of the cut to suit the size of the biscuit and adjust the fence to align the blade with the centre line.
◆ Line up the centre cutting mark on the tool with the central mark of the slot. Press the fence against the work. Start the motor and plunge the blade to make the slot. Retract the blade and allow the motor to stop before moving on.
◆ Repeat the above stages to cut the remaining slots in both pieces to be jointed.

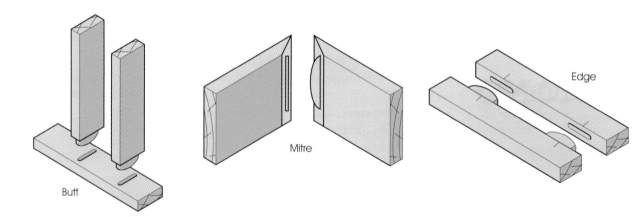

Butt   Mitre   Edge

Figure 4.28 *Use of biscuits for jointing*

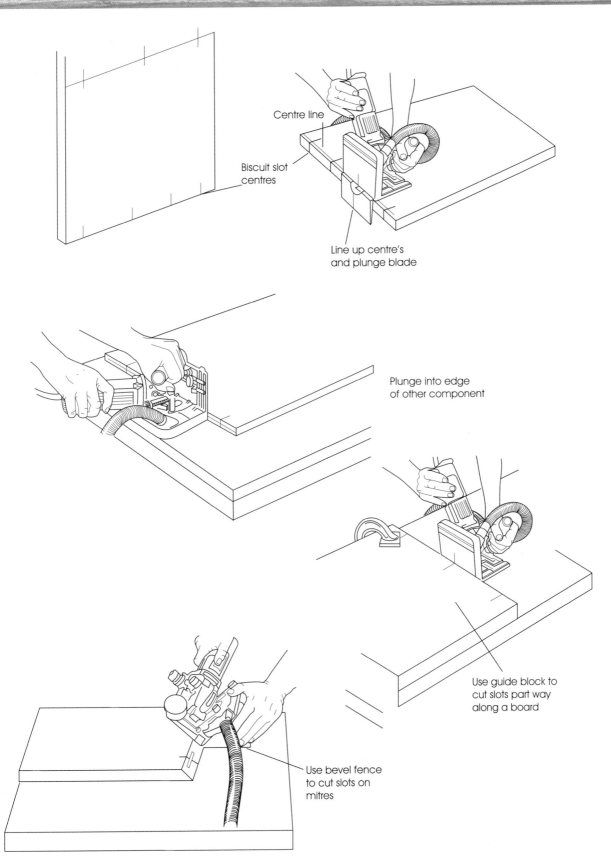

Centre line

Biscuit slot
centres

Line up centre's
and plunge blade

Plunge into edge
of other component

Use guide block to
cut slots part way
along a board

Use bevel fence
to cut slots on
mitres

Portable Powered Hand Tools

Chapter 4

Figure 4.29 *Biscuit joint preparation*

Planers and routers

Planers

The portable planer (Figure 4.30) is mainly used for edging work although it is capable of both chamfering and rebating. On site it is extremely useful for door hanging and truing up the edges of sheet material. Surfacing and cleaning up of timber can be carried out when required but it tends to leave ridges on surfaces, which are wider than the length of the cutters.

Operation of a planer

- Check that the cutters are sharp and set correctly.
- Adjust the fence to run along the edge of the work as a guide.
- Rest the front of the plane on the workpiece, ensuring that the cutters are not in contact with the timber.
- Pull the trigger and allow the cutters to gain speed.
- Move the plane forward, keeping pressure on the front knob. The depth of the cut can be altered, by rotating this knob.
- Continue planing, keeping pressure both down and up against the fence (Figure 4.31).
- When completing the cut, ease the pressure off the front knob and increase the pressure on the back. This prevents the plane tipping forward, which would make the cutter dig in when the end of the cut was reached.
- Allow the cutters to stop before putting the plane down, otherwise the plane could take off on the revolving cutters.

Rebating –

- Adjust the side fence and depth gauge to the required rebate dimensions.
- Use repeated passes to plane to the required depth, keeping the fence pressed up against the workpiece at all times.

Allow the cutters to stop before putting a plane down.

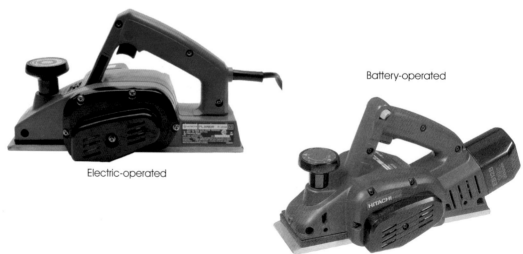

Battery-operated

Electric-operated

Figure 4.30 *Planers*

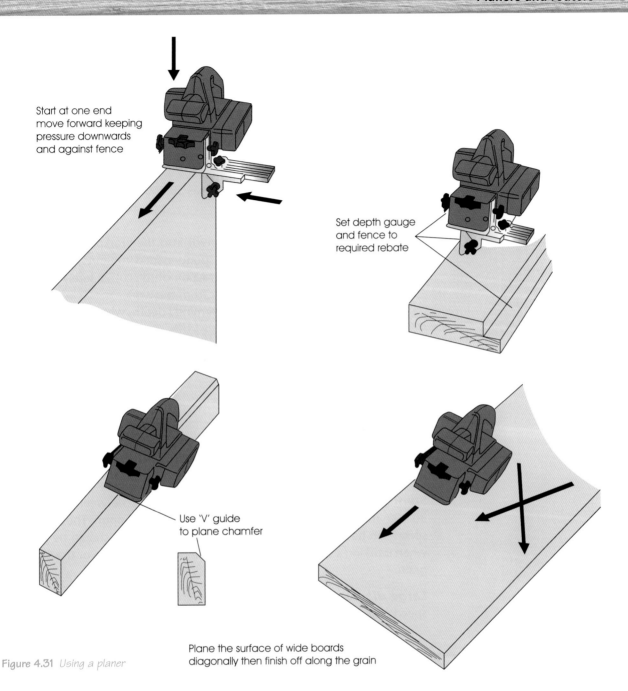

Start at one end move forward keeping pressure downwards and against fence

Set depth gauge and fence to required rebate

Use 'V' guide to plane chamfer

Plane the surface of wide boards diagonally then finish off along the grain

Figure 4.31 *Using a planer*

Chamfering –

- Use the 'V' guide on the in-feed sole plate to locate the planer on the 90° corner of the workpiece.
- Use a number of passes, to plane the required 45° chamfer.
- The side fence may be adjusted to run along the edge of the workpiece, to guide the plane as the chamfer widens, or when angles other than 45° are required.

Surfacing a wide board –

- Plane diagonally across the board in two directions, overlapping each pass.
- Plane parallel to the edge of the board using a number of overlapping passes to cover the entire board width.

Portable powered routers

The power router has largely taken the place of a wide range of moulding, grooving and rebating planes (Figure 4.32). In addition they are also used for trimming, recessing, housing, slot mortising, dovetailing, drilling and plunge cutting. Mechanically the majority of routers are similar, with the cutter (bit) mounted directly below a motor housing that is fitted with a handle on either side. The motor moves up and down on two spring-loaded columns attached to the baseplate. The router works by spinning the cutter at a very high speed, as high as 30,000 rpm.

Methods of operation

- The cutter can be plunged (lowered) into the workpiece by pressing down on the motor against the spring-loaded columns (Figure 4.33). This action also allows for the retraction of the cutter safely above the baseplate before lifting the router from the work.
- The cutter can be set at a pre-determined depth, which is then entered into the workpiece sideways.
- Alternatively, the router may be mounted upside down in a table. With the cutter protruding above the table the workpiece is fed past the cutter.

Collet capacity – The shank of router cutters fit into a tapered collet and are secured by a locking nut. Collet sizes are usually either 6 mm, 8 mm or 12 mm in diameter to suit the cutter shank. Larger routers have interchangeable collets, enabling all three sizes of cutter shanks to be used.

Spindle speed – Maximum spindle speed varies between 20,000 and 30,000 rpm, depending on the motor's power output. The maximum speed can be used for most operations. Slower speeds should be selected when working some hardwoods or those with interlocking or troublesome grain.

Large diameter moulding cutters – These have a very high peripheral speed and should be run at lower spindle speeds for safety.

Plunging router

Laminate trimmer

Figure 4.32 *Routers*

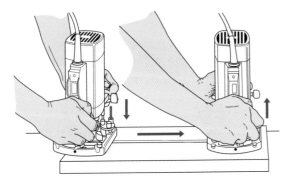

Plunge, rout and retract

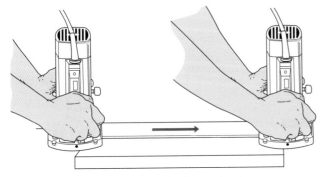

Preset cutter projection and rout

Router inverted under table, feed workpiece past the cutter

Figure 4.33 *Router operation*

Router cutters

Router cutters are available in a wide variety of sizes and shapes (Figure 4.34). They fall into a number of main categories:

Groove forming cutters – are a range of basic cutters used to make recesses both with and across the grain. They normally require guiding by a fence or guide batten.

Edge forming cutters – have a guide tip which runs against the edge of the workpiece, making the use of a fence or guide batten unnecessary. Fixed guide tips can cause the wood to burn due to friction. This can normally be removed with a finely set plane. However, cutters with a ball-bearing guide tip are available. These run against the work edge without causing the burn blemish.

Trimming cutters – there are two types available for trimming plastic laminate and solid timber edging. Both have a ball-bearing guide wheel that runs on the board material. The parallel profile is used to trim the edging strip flush and square with the board surface. The chamfered profile can be used for trimming laminate tops at an angle to the edging.

High speed steel (HSS) cutters – are suitable for most general woodworking operations, but tungsten carbide tipped (TCT) cutters, although more expensive initially, hold their edge longer, especially when working abrasive timbers, hard glue lines, sheet material and plastics. Solid tungsten carbide (STC) cutters are also available; they can be used with the same materials as TCT cutters but are considerably more expensive.

did you know?

Lower speeds should be used when routing plastics and soft metals.

safety tip

Dust extraction and PPE are particularly important when using routers.

Portable Powered Hand Tools **Chapter 4**

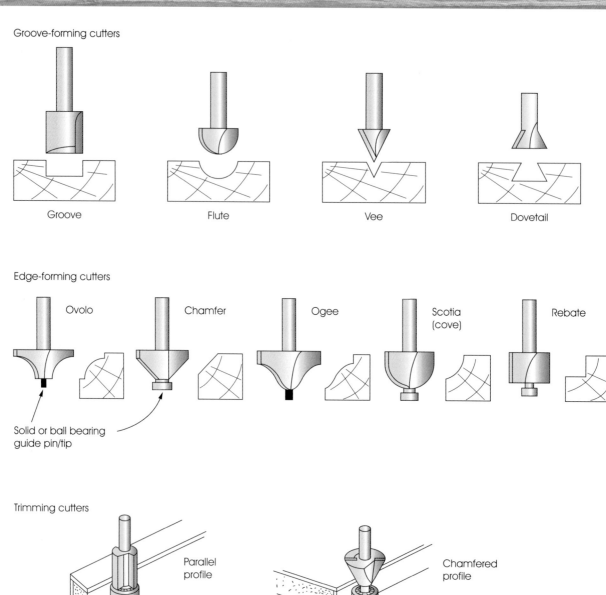

Groove-forming cutters

Groove Flute Vee Dovetail

Edge-forming cutters

Ovolo Chamfer Ogee Scotia (cove) Rebate

Solid or ball bearing guide pin/tip

Trimming cutters

Parallel profile

Chamfered profile

Ball-bearing guide

Figure 4.34 *Router cutters*

did you know?

The edges of HSS cutters can be honed by the woodworker with an oiled slipstone to extend their usage. Both TCT and STC cutters can only be resharpened with specialist grinding facilities.

Operation of a router – health and safety

When using routers fine dust particles are produced especially when working with man-made board materials. Even a short session can cover both the workshop and the operator with a cloud of lung-clogging dust. Most routers can be equipped with a clear plastic hood that fits over the base to enclose the cutter. The hood is connected by a flexible hose to a vacuum extraction unit, which collects the dust and larger waste material as it leaves the cutter. It is strongly recommended that this facility is always used with the router even for short periods of working. The hood must be used in addition to the wearing of standard personal protective equipment: ear muffs, dust masks and goggles for yourself and anyone working near you.

Methods of use

For most operations, once having set up the router, clamped the workpiece in position and connected it to the power supply, two alternative methods may be used depending on the type of router and the work in hand.

Plunging technique

◆ Place the router on the workpiece ensuring that the cutter is clear. Taking a firm grip of the router, start the motor and allow it to obtain maximum speed. Plunge the router to the pre-set depth and lock. For deep cuts, several passes will have to be made to achieve the required depth (Figure 4.35).
◆ Applying a firm downward pressure, move the router steadily forwards. Routing too slowly will cause the cutter to overheat, resulting in a poor finish and leaving burn marks on the work.
◆ At the end of the cut, release the lock, which will cause the cutter to spring up clear of the work. Switch off the motor and allow the cutter to stop rotating before moving the router. It is most important that the router is fed in the opposite direction to the rotation of the cutter. The cutter rotates clockwise when viewed from the top of the router (Figure 4.36).

Pre-set depth technique

◆ Set the router cutter to the required projection below the baseplate. Deep cuts will have to be done in several stages.
◆ Rest part of the router baseplate on the workpiece. Ensure the projecting cutter is clear of the work. Start the motor and allow it to obtain maximum speed.
◆ Applying a firm downward pressure and side pressure against a guide batten or fence, move the router steadily forwards to make the cut.
◆ When the cutter emerges at the other end, switch off and allow the cutter to stop rotating before lifting the baseplate off the workpiece.

Housing or trenching using guide battens

◆ Clamp a batten across the workpiece. Ensure that the batten projects beyond both edges (Figure 4.37). When positioning the batten an allowance must be made equal to the distance between the cutting edge of the bit and the baseplate. For housings wider than the cutter two battens are used.
◆ Lower and lock the router plunge mechanism to the required cutter projection.
◆ Rest the router baseplate on the workpiece with the cutter clear of the work.
◆ Switch on and allow the motor to reach its working speed.
◆ Keeping the router baseplate pressed against the batten, feed the router into and through the workpiece at a steady rate until the cut is complete. For wide housings the first cut is made against the left-hand batten. Then move the router across to the right-hand batten to complete the cut. Using this method the rotation of the cutter helps to hold the router base against the guide battens.
◆ Switch off with the cutter just clear of the work. Allow the cutter to stop before lifting the baseplate off the workpiece.

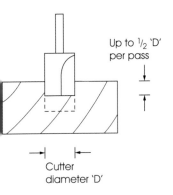

Up to $\frac{1}{2}$ 'D' per pass

Cutter diameter 'D'

Figure 4.35 Deep cuts are made in several passes

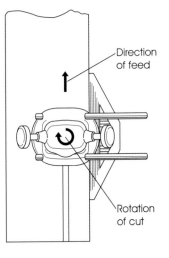

Direction of feed

Rotation of cut

Figure 4.36 Direction of feed and cutter rotation

did you know?

Rout deep cuts in several passes to avoid damaging the workpiece cutter and motor.

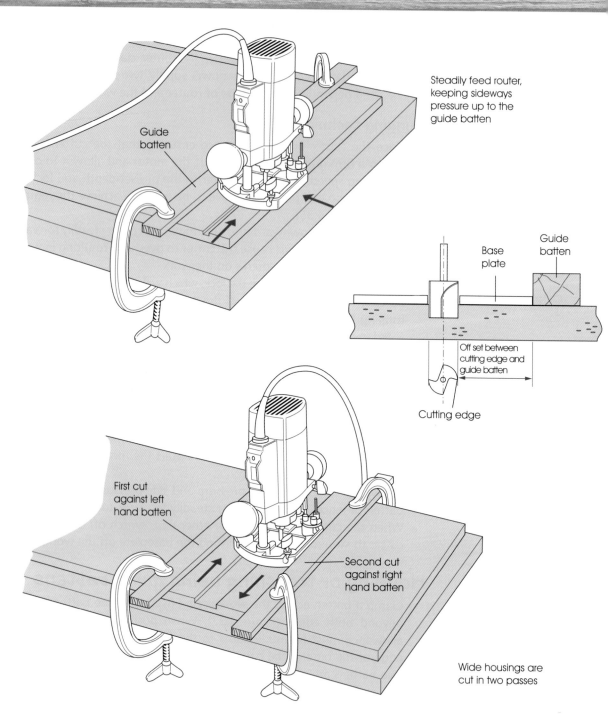

Steadily feed router, keeping sideways pressure up to the guide batten

Guide batten

Base plate

Guide batten

Off set between cutting edge and guide batten

Cutting edge

First cut against left hand batten

Second cut against right hand batten

Wide housings are cut in two passes

Figure 4.37 *Housing or trenching with guide battens*

Grooving using the fence

Most grooves are required to be parallel with and close to the edge of the workpiece. The bolt-on side fence is ideal for this operation.

◆ With the router unplugged, rest the router on the workpiece with the cutting edge of the cutter aligned with the edge of the groove marked on the surface.
◆ Adjust the side fence up to the edge of the workpiece and tighten its thumbscrews (Figure 4.38).
◆ Cut the groove using either the pre-set depth or plunge technique. In either case keep side pressure on the fence throughout the cut.

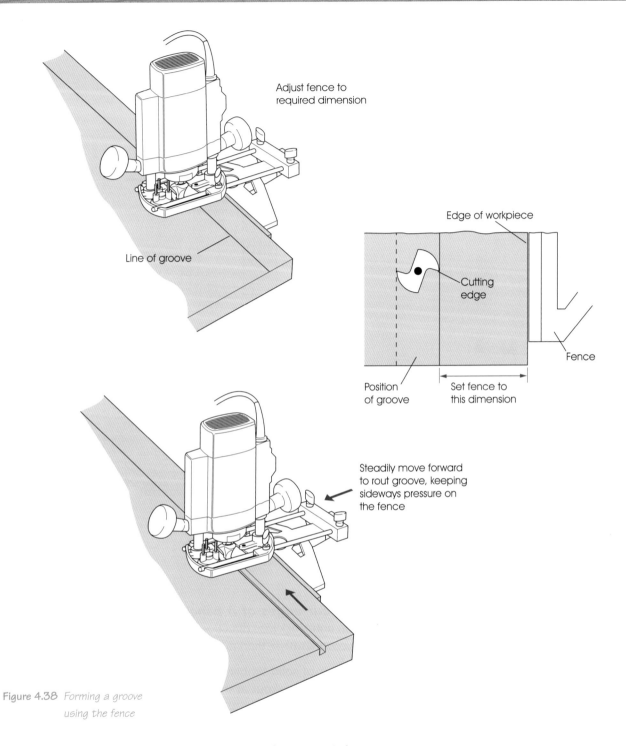

Adjust fence to
required dimension

Line of groove

Edge of workpiece

Cutting
edge

Fence

Position
of groove

Set fence to
this dimension

Steadily move forward
to rout groove, keeping
sideways pressure on
the fence

Figure 4.38 *Forming a groove
using the fence*

Rebating with a straight cutter

Rebates are best cut using an edge-forming cutter with a guide pin or ball-bearing guide. However, they can be cut with a straight cutter and side fence, using a similar procedure as that used for cutting grooves (Figure 4.39).

Moulding with an edge-forming cutter

Moulding or rebating the edge of a workpiece is achieved using an edge-forming cutter with a guide pin or ball-bearing guide (Figure 4.40).

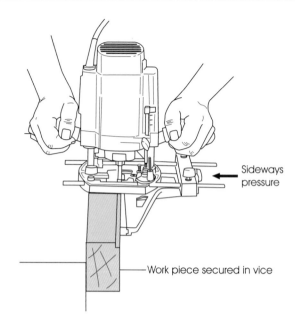

Sideways pressure

Work piece secured in vice

Figure 4.39 *Rebating with a straight cutter*

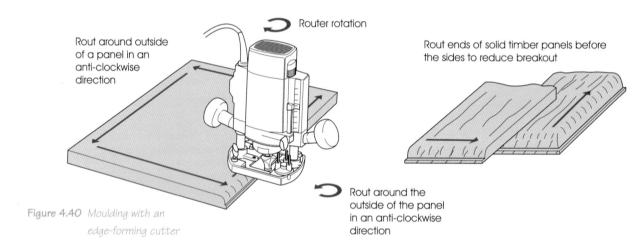

Rout around outside of a panel in an anti-clockwise direction

Router rotation

Rout ends of solid timber panels before the sides to reduce breakout

Rout around the outside of the panel in an anti-clockwise direction

Figure 4.40 *Moulding with an edge-forming cutter*

Moulding all four edges of a panel – Work the outer edge of the workpiece in an anti-clockwise direction, as this ensures the cutter rotation pulls it into the work rather than pushing it off.

Where the panel is of solid timber, mould the end grain before the sides. This ensures that any breakout at the far edge of the cut is removed when the sides are moulded.

Moulding the inside of an assembled frame – Use the same technique as moulding the outer edge of a panel, but move the router in a clockwise direction to keep the cutter in contact with the workpiece (Figure 4.41).

Rounded corners will be left by the cutter; for mouldings these can be an additional feature. Rebates for glazing will require squaring up with a chisel.

Recessing with a template

This operation requires the fitting of a template guide to the baseplate of the router and a template of the recess required (Figure 4.42).

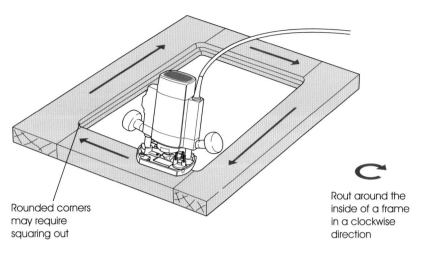

Rounded corners
may require
squaring out

Rout around the
inside of a frame
in a clockwise
direction

Figure 4.41 *Moulding the inside of an assembled frame*

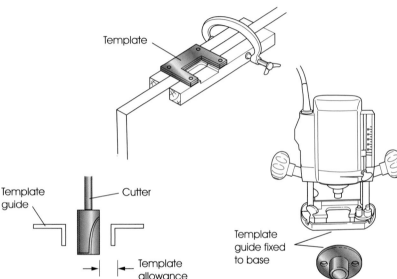

Template

Template
guide

Cutter

Template
allowance

Template
guide fixed
to base

Figure 4.42 *Recessing with a template*

When making the template an allowance must be made all round equal to the distance between the cutting edge of the bit and the outside edge of the template guide.

- Fix the template in the required position.
- Place the router base on the template. Taking a firm grip on the router, start the motor and allow it to attain its working speed. Plunge the router to the pre-set depth and lock.
- Applying a firm downward pressure, move the router around the edge of the template before working the centre. It is most important to feed the router in the opposite direction to the rotation of the cutter.
- On completion, retract the cutter, switch off the motor and allow the cutter to stop rotating before putting the router down. Rounded corners will be left by the cutter which can easily be squared up with a sharp chisel.

Cutting circles

This is achieved with the aid of a trammel bar, which is fixed to the baseplate at one end and has a centre pin at the other (Figure 4.43).

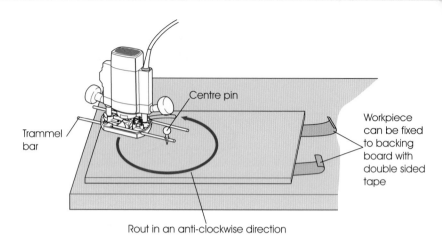

Figure 4.43 *Routing circles*

The router is rotated in an anti-clockwise direction about the centre pin to create either a circular hole in the workpiece or a circular disk.

Cutting with a template

Routing shapes against a template is the ideal way of cutting shapes or making a number of identical components (Figure 4.44). Templates are best cut from MDF, since accuracy is essential as any minor defect will be reproduced in the finished components.

A template guide is fixed to the baseplate to run against the edge of the template. Therefore an allowance has to be made in the template equal to the distance between the cutting edge and the outer edge of the template guide.

Templates can either be pinned to the workpiece or fixed to it using double-sided tape, depending on the intended finish.

Again, on outside edges rout in an anti-clockwise direction to keep the template guide in contact.

Laminate trimming

Small routers for single-hand operation are manufactured specifically for laminate trimming, although it is possible to fit a ball-bearing guided laminate trimmer to a standard plunging router.

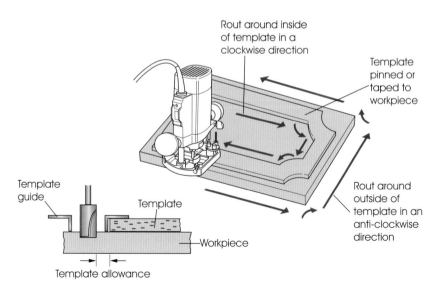

Figure 4.44 *Routing shapes with a template*

Cutting joints

The router can be used to cut a wide range of joints.

Rebated and grooved joints

Lapped, bare-faced tongues and halving joints can all be cut using a similar technique (Figure 4.45).

- ◆ Clamp the components to be jointed side-by-side to the bench top.
- ◆ Use a straight cutter in the router, with the baseplate against a guide batten to make the cut.
- ◆ Halving joints will require several passes to cut the required width.

Tongue and grooved joints

- ◆ Set up the router with a straight cutter and guide fence.
- ◆ Secure in the vice the piece to receive the tongue between two scrap pieces of timber (Figure 4.46). This gives a wider flat surface to give better support to the baseplate.
- ◆ Pass the router along both sides of the piece to form the tongue.
- ◆ Secure the piece to be grooved in the vice, again between two pieces.
- ◆ Reset the guide fence to position the cutter in the centre and pass the router along the piece to form a matching groove.

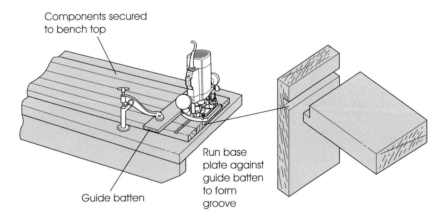

Components secured to bench top

Guide batten

Run base plate against guide batten to form groove

Figure 4.45 *Routing grooved joints*

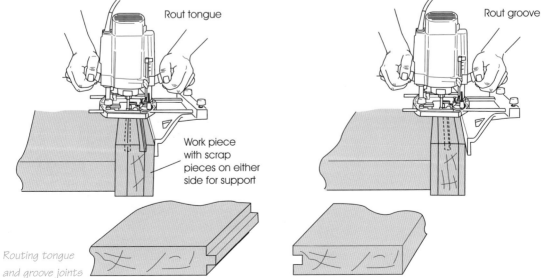

Rout tongue

Rout groove

Work piece with scrap pieces on either side for support

Figure 4.46 *Routing tongue and groove joints*

Portable Powered Hand Tools

Chapter 4

Mortise and tenon joints

◆ Set up the router with the required mortise width, straight cutter and guide fence.
◆ Secure the piece in the vice using scrap timber on either side to increase the supporting width (Figure 4.47).
◆ Cut the mortise in stages as a short stopped groove, gradually increasing the depth of cut with each pass. Stub mortises can be routed from one edge, though mortises may require routing from both edges.
◆ Pencil lines can be used as a guide to the positions for plunging and retracting the router, but for greater accuracy pin stop-blocks to one of the scrap pieces.

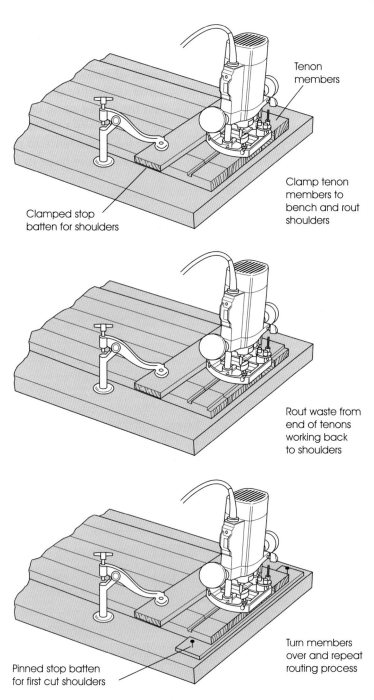

Tenon members

Clamped stop batten for shoulders

Clamp tenon members to bench and rout shoulders

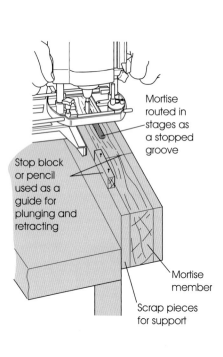

Mortise routed in stages as a stopped groove

Stop block or pencil used as a guide for plunging and retracting

Mortise member

Scrap pieces for support

Rout waste from end of tenons working back to shoulders

Turn members over and repeat routing process

Pinned stop batten for first cut shoulders

Figure 4.47 *Routing a mortise and tenon joint*

- Use a chisel to square up the rounded ends of the mortise.
- Lay the pieces to be tenoned side-by-side on the bench. Secure using a clamped guide batten parallel with, but offset from, the shoulder.
- Set the cutter projection to the required depth and cut all the shoulders in one operation.
- Remove the remaining waste freehand starting at the ends of the tenon for support, gradually working back towards the shoulder.
- Turn pieces over, butt the cut shoulders against a stop batten pinned to the bench.
- Secure with a clamp and batten and repeat the previous stage to complete the tenons.

Dovetail joints

These can be cut with a router that is set up with a dovetail cutter and a manufacturer's dovetail jig (Figure 4.48). The setting up and use of these vary; detailed instructions are supplied and must be followed.

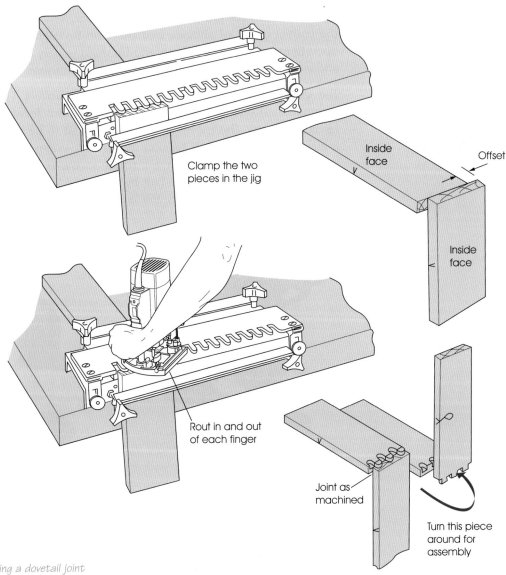

Clamp the two pieces in the jig

Inside face

Offset

Inside face

Rout in and out of each finger

Joint as machined

Turn this piece around for assembly

Figure 4.48 *Routing a dovetail joint*

◆ Clamp the two pieces to be joined in the jig; these should be inside out and slightly offset.
◆ Set the dovetail cutter to the required depth and cut the joint, passing the router in and out of each finger of the jig template.
◆ Remove both pieces from the jig and turn them around ready for assembly.

Sanders

The use of power sanders has taken most of the effort out of the process of shaping and finishing work. However, even the finest of finishing sanders still leave minute surface scratches, which may require a final period of hand sanding to remove, especially when the surface is to receive a polished or varnished finish. Types of sanders are shown in Figure 4.49.

Belt sanders

These are used for jobs requiring rapid stock removal. When fitted with the correct grade of abrasive belt they can be used for a wide range of operations, such as smoothing and finishing joinery items, block flooring and even the removal of old paint and varnish finishes.

The sanding or abrasive belt is fitted over two rollers. The front roller is spring-loaded and can be moved backwards and forwards by the belt-tensioning lever. This movement allows the belt to be changed easily and it also applies the correct tension to the belt. When changing the belt it is necessary to ensure that it will rotate in the correct direction. This is indicated on the inside of the belt by an arrow. If the belt is inadvertently put on the wrong way round, the lap joint, which runs diagonally across

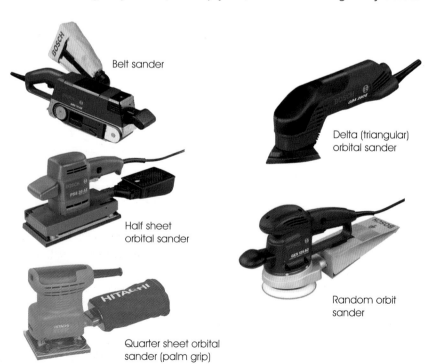

Belt sander

Delta (triangular) orbital sander

Half sheet orbital sander

Random orbit sander

Quarter sheet orbital sander (palm grip)

Figure 4.49 *Types of power sander*

Chapter 4   Portable Powered Hand Tools

the belt, will tend to peel. This could result in the belt breaking, with possible damage to the work surface. To keep the belt running central on the rollers, there is a tracking control knob on the sander, which adjusts the front roller by tilting it either to the left or right as required. The tracking is adjusted by turning the sander bottom upwards, with the belt running, and rotating the tracking knob until the belt runs evenly in the centre without deviating to either side.

Orbital sanders

Also known as the finishing sander as it is mainly used for fine finishing work. The sander's base has a 3 mm orbit which operates at 12,000 rpm.

Various grades of abrasive paper may be clipped to the sander base. It is best to start off with a coarse grade to remove any high spots or roughness and follow on with finer grades until the required finish is obtained. However, where the surface of the timber has machine marks or there is a definite difference in the levels of adjacent material, the surface should be levelled by planing before any sanding is commenced.

The baseplate size is based on a proportion of the standard size sheet used for handwork. Larger machines are termed half- and third-sheet sanders. Smaller, single-handed palm grip machines use a quarter sheet. Delta sanders have a small triangular baseplate for finishing in tight corners.

Random orbit sanders – work with an off-centre (eccentric) orbital action plus a simultaneous rotation, to give a surface that is virtually scratch free. The circular rubber backing pad is flexible, to enable convex and concave surfaces to be finished.

Abrasive discs – are self-adhesive, with either a peel-off backing or Velcro backing, for efficient fitting and replacement.

Most sanders are fitted with a dust collection bag, or provision for connecting to a vacuum extract system. Some types require the use of perforated abrasive sheets to allow the dust and fine abrasive particles produced to be sucked up.

Operation of sanders

- ◆ Always start the sander before bringing it into contact with the work surface and remove it from the work surface before switching off. This is because slow moving abrasive particles will deeply scratch the work surface.
- ◆ Do not press down on a sander in use (Figure 4.50). The weight of the machine itself is sufficient pressure.
- ◆ For best results lightly guide the sander over the surface with parallel overlapping strokes in line with the grain.
- ◆ Always use the abrasive belts and sheets specifically recommended by the manufacturer for the particular model as makeshift belts and sheets are inefficient and dangerous.
- ◆ Always use the dust collecting bag or extraction system where fitted. Always wear a dust mask as inhaling the dust from many species of wood causes coughing, sneezing and runny eyes and nose and possible allergic reactions.

did you know?

Always use the dust collection/extraction system, because as well as creating a health hazard, the dust and abrasive particles, if not removed, will clog the sheet and cause further scratches to the work.

did you know?

Excessive pressure causes clogging of the abrasive material and overheating of the motor.

Portable Powered Hand Tools

Chapter 4

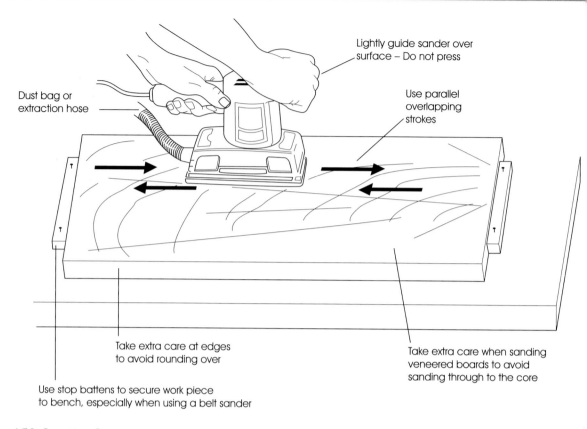

Lightly guide sander over surface – Do not press

Dust bag or extraction hose

Use parallel overlapping strokes

Take extra care at edges to avoid rounding over

Take extra care when sanding veneered boards to avoid sanding through to the core

Use stop battens to secure work piece to bench, especially when using a belt sander

Figure 4.50 *Operation of sanders*

Automatic pin drivers

Various pin gun nailers and staplers are available to suit a wide range of fixings (Figure 4.51). The smaller drivers are usually operated by spring power, the larger ones by compressed air, although a limited range of gas-operated nailers and electric tackers are available.

Operation of automatic drivers will vary depending on the type being used but general points to note are:

◆ Do not operate the trigger until the baseplate is in contact with the fixing surface.
◆ Keep fingers clear of the baseplate.
◆ Maintain a firm pressure with the fixing surface during use. Failure to do so can result in kickback of the tool and ricochet of the nail or staple.
◆ On tools with a variable power adjuster, carry out trial fixings at a low setting and gradually increase until the required penetration is achieved.

safety tip

Keep fingers clear of baseplate. Nails can deflect and come out of the side of the workpiece.

Chapter 4 Portable Powered Hand Tools

Hand-held spring stapler

Compressed air
pin gun nailer/stapler

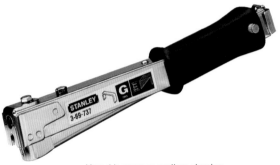

Hand hammer-action stapler

Cordless gas-operated
nail gun

Figure 4.51 *Automatic drivers*

Portable Powered Hand Tools Chapter 4

activity

You are asked by the workshop foreman where you are working to help plan the company's new maintenance of portable powered hand tools.

Write a memo to the workshop foreman listing the procedures the company and the employees should follow to ensure the safe working and maintenance of portable powered handtools.

1. Produce a sketch to show the correct direction of feed for a router in relation to the rotation of the cutter.

2. State the purpose of using an RCD device.

3. Describe the procedure for plunge cutting with a circular saw.

4. In order to reach the workplace an extension lead is required. State where this should be positioned when using a 110-volt step down transformer.

5. List FOUR safety precautions to be followed when using any portable powered tool.

6. Describe the action to be taken when you suspect a portable power tool is not working correctly or its safety is suspect.

7. Explain what the torque control is for on a battery-operated drill/driver.

8. Name the mould that the router cutter illustrated below will make on the edge of the workpiece.

9. Which TWO of the following statements are incorrect?
   1) Routing too slowly causes burn marks
   2) Belt sanders are good for finishing
   3) All power tools must be started and stopped under load
   4) 110 V tools are considered safer than 240 V tools

10. Name TWO portable powered hand tools that can be used for rebating.

Installing Frames and Linings

This chapter is intended to provide the reader with an overview of installing frames and linings. Its contents are assessed in Install Frames and Linings NVQ unit VR 05.

In this chapter you will cover the following range of topics:

◆ Frame and lining terminology
◆ Frames
◆ Linings

What's required in VR 05?

To successfully complete this unit you will be required to demonstrate your skill and knowledge of the following processes:

◆ Interpreting information
◆ Adopting safe and healthy working practices
◆ Selecting materials, components and equipment
◆ Preparing and fixing frames and linings

You will be required practically to:

◆ Prepare and fix the following:
  ▸ door frames;
  ▸ door linings;
  ▸ hatch linings;
  ▸ window frames;
  ▸ window boards.
◆ Use hand and portable power tools.
◆ Communicate with other team members.
◆ Use access equipment.
◆ Work at height.

Frame and lining terminology

Frame – an assembly of components to form an item of joinery, such as a door or window; a structural framework of columns and beams or panels in steel-reinforced concrete or timber.

Lining – the thin covering to door or window reveals; sheet material used to cover wall surfaces.

Door frame – the surround on which an external door or internal door is hung, consisting of two jambs, a head and sometimes a threshold and transom; normally with stuck-on solid stops and of a bigger section than door linings (Figure 5.1).

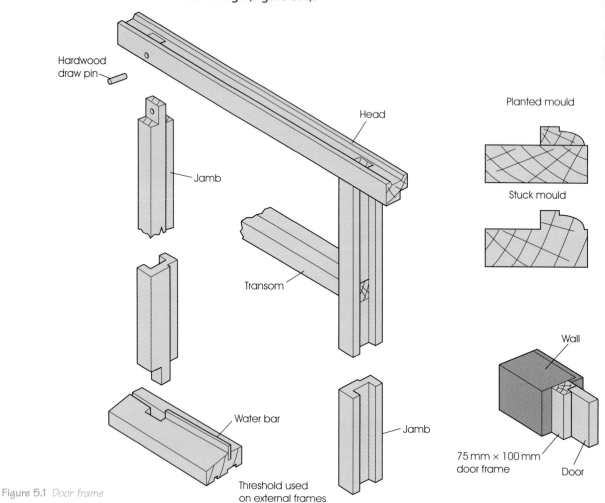

Figure 5.1 *Door frame*

Door lining – the surround on which mainly internal doors are hung, normally of a thinner section than door frames and often with planted stops. The main difference between door frames and door linings is that linings cover the full width of the reveal in which they are fixed from wall surface to wall surface, whereas frames do not (Figure 5.2).

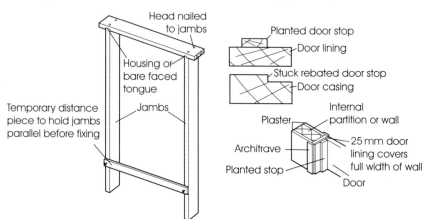

Figure 5.2 *Door lining*

The frame material is used as the first classification of windows, e.g. timber or plastic windows, followed by whether or not they open.

Window frame – the part of a window that is fixed into the wall opening, which receives the casements or sashes. Windows that do not open are termed fixed or direct glazing; others are termed by their method of opening and whether or not they project from the main building line (Figure 5.3).

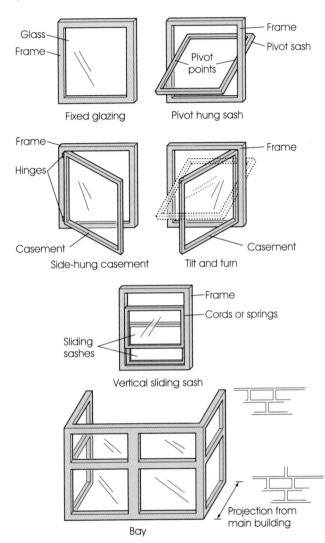

Figure 5.3 *Classification of windows*

Installing Frames and Linings

Chapter 5

Frames

Frames can be either 'built-in' or 'fixed-in'.

◆ Built-in frames are fixed into a wall or other element by bedding in mortar and surrounded with the walling components.
◆ Fixed-in frames are inserted into a ready formed opening after the main building process.

Built-in frames

The majority of frames are built-in by the bricklayer as the brickwork proceeds. Prior to this the frame has to be accurately positioned, plumbed, levelled and temporarily strutted by the carpenter.

The foot of the door frame jambs, in the absence of a threshold, are held in position by galvanised metal dowels which are drilled into the end of the jambs and are grouted into the concrete as (Figure 5.4). Temporary struts are used to hold the frame upright. Temporary braces and distance pieces are fixed to the frame, in order to keep it square and the jambs parallel during the 'building-in' process.

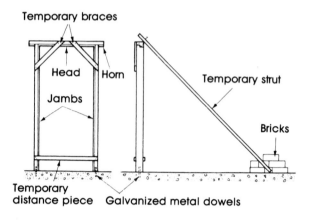

Figure 5.4 'Building in' a frame

The frame's head should be checked for level and packed up as required; doorframes with thresholds and the sill of windows are normally bedded level using bricklayer's mortar (Figure 5.5).

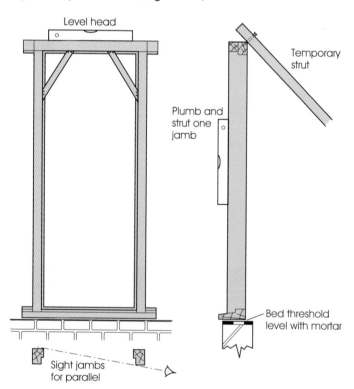

Figure 5.5 Building in a frame with threshold

Jambs should be plumbed from the face. It is standard practice to plumb and fix the first using a spirit level. The other is then sighted parallel: stand to the side of the frame, close one eye, sight the edge of the plumbed jamb with the edge of the other and adjust if required until both jambs are parallel.

Figure 5.6 *Attaching frame cramps*

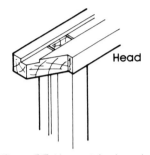

Figure 5.7 *Horn cut back ready for 'building in'*

As the brickwork proceeds, galvanised metal frame cramps should be screwed to the back of the jambs and built into the brickwork (Figure 5.6). Three or four cramps should be evenly spaced up each jamb.

The horns of the frame should be cut back as shown before 'building in', rather than cut off flush (Figure 5.7).

The horns will then help to fix the frame in position. After trimming the horns it is essential that the cut ends are treated with preservative in order to reduce the possibility of timber rot. This can be carried out by applying two brush flood coats of preservative.

Storey height frames may be used for internal door openings in thin blockwork partitions. The jambs and head, which make up the frame, are grooved out on their back face to receive the building blocks. The storey frame should be fixed in position, at the bottom to the wall plate and at the top to the joists, before the blocks are built up (Figure 5.8). The jambs above the head are cut back to finish flush with the blockwork. As with other frames, one jamb should be fixed plumb and the other sighted to it.

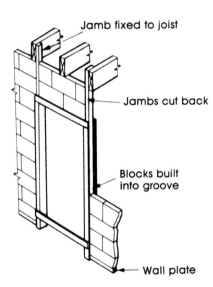

Figure 5.8 *'Building in' a storey height frame*

Fixed-in frames

These are installed into prepared openings. This applies mainly to expensive hardwood and plastic frames, in order to protect them from possible damage or discoloration during the building process. In addition, frames that are not available during the building process, or replacement frames, will have to be fixed-in at a later stage.

Horns on fixed-in frames are not required as a fixing and should be sawn off flush with the back of the jambs. Remember to preservative treat the cut ends.

Place the frame in the prepared opening, temporarily holding it in position with the aid of folding wedges. Check the head or sill for level and adjust the wedges as required. Plumb one jamb and 'sight in' the other; adjust the wedges if required (Figures 5.9 and 5.10).

Installing Frames and Linings   **Chapter 5**

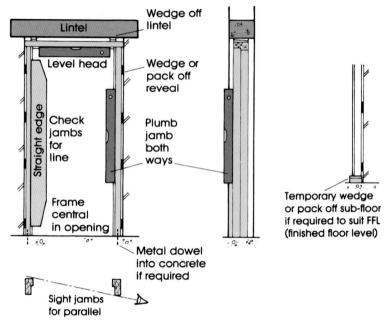

Figure 5.9 *Positioning 'fixed-in' door frames*

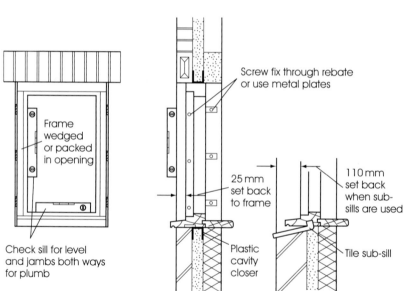

Figure 5.10 *Positioning 'fixed-in' window frames*

Figure 5.11 shows how to fix the frame to the wall using one of the following methods:

Nailing – using cut nails into the blockwork or brickwork mortar joints or masonry nails into the actual brick.

Screwing – plastic plugs and screws through the jamb. Screwheads in softwood frames may be countersunk below the surface and filled. Screwheads in hardwood frames should be concealed by counter boring and pelleting.

Metal plates – used as fixing lugs screwed at intervals to the back of the jamb before the frame is put in the opening. The lugs are screwed and plugged to the brick or block reveals.

Frame anchors – proprietary fixing consisting of a metal or plastic sleeve and matching screw. The jamb and reveal are drilled out to suit the sleeve, which is inserted in position and screwed up tight.

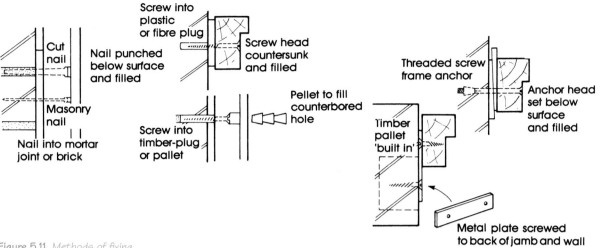

Figure 5.11 *Methods of fixing*

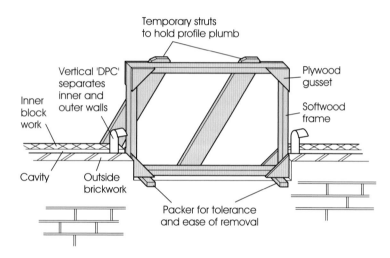

Figure 5.12 *Profile frame for windows that are to be fixed at a later stage*

did you know?

Any slight unevenness in the face bead can be smoothed off using a small paintbrush dipped in water.

Profiles

Where window frames are fixed in at a later stage the carpenter is often required to make a timber template or profile frame. These are made up either on-site or in the workshop using 50 mm × 75 mm softwood and plywood corner gussets (Figure 5.12).

The profile is made about 6 mm wider than the actual window as a fixing tolerance. The vertical tolerance is achieved by setting it on packing blocks to aid removal. After the wall has set, profiles may be removed, leaving the prepared opening ready for the frame.

Weather sealing of external frames

All external frames should have the gap between them and the wall sealed with a silicone or acrylic sealant prior to occupation of the building. This is to prevent the penetration of wind, rain and insects into the gap and possibly into the building.

Frame sealants are normally supplied in cartridges and applied using a skeleton gun. The sealant should be forced as far back into the gap as possible, finally finishing off with a narrow angled face bead (Figure 5.13).

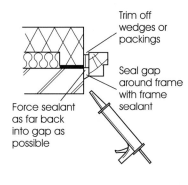

Trim off wedges or packings

Seal gap around frame with frame sealant

Force sealant as far back into gap as possible

Figure 5.13 *Weather sealing of external frames*

Window boards

These are normally used to finish the top of the wall internally at the window sill level. They may be formed from solid timber, plywood, MDF or plastic and should be fitted at the first fixing stage before the internal walls have been plastered. Where the window board fits up to a timber window, it is normally tongued into the sill groove. If the window is metal or plastic (PVCu) the window board is only butted to the sill and the gap is filled with caulking or covered with a small bead (Figure 5.14).

Many window boards have a groove on the underside; this provides a finish for the plaster and masks any gap, which may result later from shrinkage. Alternatively, a cover mould may be used to mask the joint.

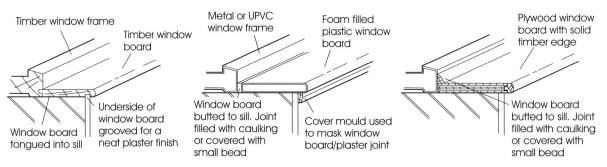

Timber window frame

Timber window board

Underside of window board grooved for a neat plaster finish

Window board tongued into sill

Metal or UPVC window frame

Window board butted to sill. Joint filled with caulking or covered with small bead

Foam filled plastic window board

Cover mould used to mask window board/plaster joint

Plywood window board with solid timber edge

Window board butted to sill. Joint filled with caulking or covered with small bead

Figure 5.14 *Window board details*

Cutting and fixing window boards (Figure 5.15)

◆ Cut a length of board about 100 mm longer than the sill.
◆ Mark out each end and cut out a portion of the board to fit the window reveal.
◆ Using a plane, return the front edge nosing profile around the ends, then finish with an abrasive paper.
◆ Use an off-cut of window board and a boat level to check for front to back level. Packing may be required under the board where it is not level; proprietary plastic shims available in a range of thicknesses are the best. However, a piece of hardboard or several pieces of damp-proof material will also do the job. These should be positioned about 50 mm from each end, with intermediates spaced at about 450 mm centres.
◆ Place the window board in position and fix the front edge to the wall, using either cut nails, masonry nails or plugs, screws and pellets. Fixings should preferably go through any packers so the board is not pulled down out of line and also to ensure they are not misplaced later.

Alternative fixing methods

A batten or ground may be pre-fixed to the wall at the correct level for the window board, using cut or masonry nails; the window board in turn is fixed directly to the ground using ovals nails. Plastic and other pre-finished window boards may be bedded and fixed to the wall using beads of a gap-filling, 'no nails' type adhesive, or frame anchors screw-fixed to the underside of the window board and wall face.

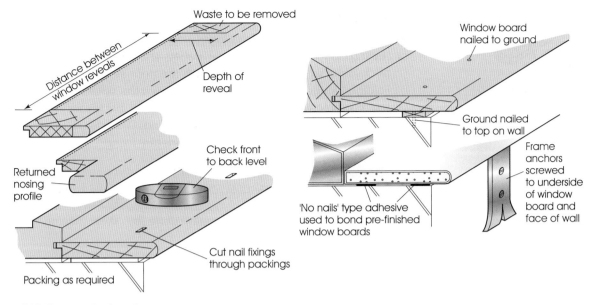

Figure 5.15 *Fixing window boards*

Linings

Plain linings – consist of two plain jambs and a plain head joined together using a bare-faced tongue and housing. The pinned stop is fixed around the lining after the door has been hung (Figure 5.16).

Rebated linings – are used for better quality work. They consist of two rebated jambs and a rebated head. The rebate must be the correct width so that when the door is hung it finishes flush with the edges of the lining (Figure 5.17).

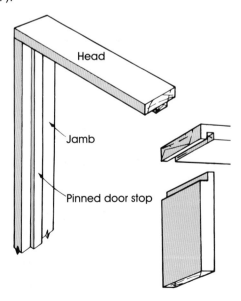

Figure 5.16 *Plain lining*

Fixing linings

The opening in the wall to receive the lining is normally formed while the wall is being built; the lining is fixed at a later stage.

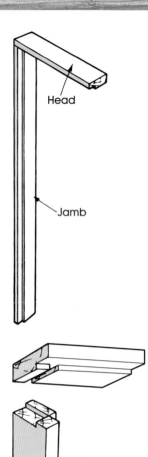

Figure 5.17 *Rebated lining*

Fixings may be:

◆ Nailing to twisted wooden plugs (see sequence of operations)
◆ Nailing and screwing to timber pallets that have been built into the brick joints by the bricklayer, or into the door stud of a stud partition. Folding wedges are used as packings down the sides of the jambs.
◆ Nailing directly into the blockwork reveal or brickwork mortar joint. Folding wedges will be required as packing.
◆ Using plugs and screws or other proprietary fixing.

Sequence of operations to fix a lining

This procedure uses twisted timber plugs (Figure 5.20):

1. Assemble lining. This is normally done by skew nailing through the head into the jambs.

2. Fix a distance piece near the bottom of the jambs, and when required, diagonal braces at the head.

3. Rake out brickwork joints and plug (Figure 5.21).

4. There should be at least four fixing points per jamb. Omit this stage if the bricklayer has 'built-in' wooden pallets or pads into the brickwork.

5. Offer lining into opening and mark where the plugs need to be trimmed. The plugs should project equally from both reveals.

6. Cut the plugs and check the distance with a width rod. The ends of the plugs should be in vertical alignment. Check with a straight edge and spirit level.

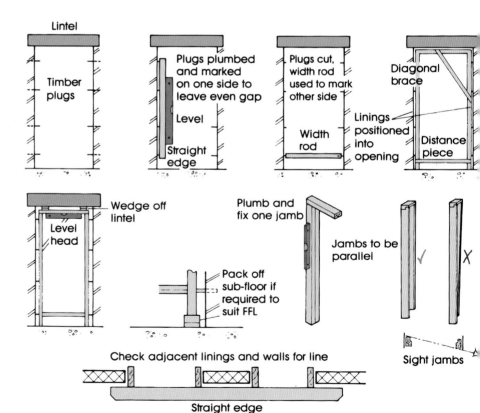

Figure 5.20 *Fixing a lining*

7. Fix the lining plumb and central in the opening by nailing or screwing through the jambs into the plugs. Before finally fixing, check head for level, wedge off lintel, ensure the lining is out of wind: check by sighting through the jambs. When fixing to unplastered walls, check adjacent linings and wall surfaces are lined up.

8. Ensure lining jambs are packed up off a concrete sub floor if required to suit the finished floor level (FFL).

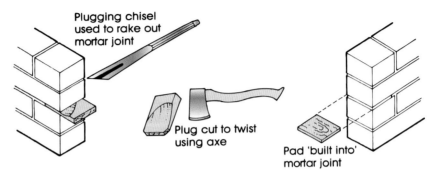

Figure 5.21 *Plugs or pads used for fixing lining*

Refer to the floor plans and schedules shown in Figures 5.22 and 5.23. Complete the table below to identify the door linings required for the house.

Door number	Location	Section of timber used for the lining or frame	Door size
D1			
D2			
D3			
D4			
D5			
D6			
D7			
D8			
D9			
D10			

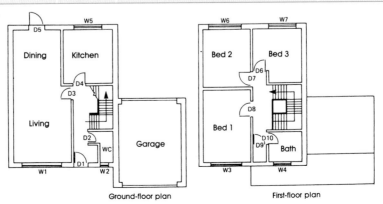

Figure 5.22 *Floor plans*

Figure 5.23 *Door schedules*

Door Schedule / frames / lining — BBS DESIGN — JOB TITLE: PLOT 3 Hilltop Road

Description	D1	D2	D3	D4	D5	D6	D7	D8	D9	D10	NOTES
Frames											
75 mm × 100 mm (outward opening)					●						
75 mm × 100 mm (inward opening)	●										
Linings											
38 mm × 125 mm		●	●	●							
38 mm × 100 mm						●	●	●	●	●	
Shape											
Rebated stop	●				●						
Planted stop		●	●	●		●	●	●	●	●	
Transom		●	●	●	●	●	●	●	●	●	
Sill	●				●						
Material											
Hardwood	●										
Softwood		●	●	●	●	●	●	●	●	●	
Fanlight infill											
6 mm tempered safety glass											
clear											
obscured		●							●		
6 mm plywood								●			

JOB NO. | DRAWING NO. | SCALE | DATE | DRAWN | CHECKED

Door Schedule / doors — BBS DESIGN — JOB TITLE: PLOT 3 Hilltop Road

Description	D1	D2	D3	D4	D5	D6	D7	D8	D9	D10	NOTES
Type (see range)											
External glazed A1					●						
External panelled A2	●										
Internal flush B1									●		
Internal flush B2		●				●	●	●		●	
Internal glazed B3			●	●							
Size											
813 mm × 2032 mm × 44 mm	●				●						
762 mm × 1981 mm × 35 mm		●	●	●		●	●	●		●	
610 mm × 1981 mm × 35 mm									●		
Material											
Hardwood	●										
Softwood			●	●	●						
Plywood/polished		●									
plywood/painted						●	●	●	●	●	
Infill											
6 mm tempered safety glass											
clear			●	●	●						
obscured	●										

JOB NO. | DRAWING NO. | SCALE | DATE | DRAWN | CHECKED

1. Produce a sketch to show the difference between a door frame and a door lining.

2. Explain the difference between built-in and fixed-in frames and state an occasion where EACH might be used.

3. Explain the reason why a steel dowel may be included in the base or foot of newel post and doorframe jambs.

4. State the reason why the sawn ends of timber are treated with preservative.

5. Explain how to 'sight-in' the jambs of a frame.

6. Explain the purpose of distance pieces, used with frames and linings.

7. List THREE forms of fixing that can be used to secure frames or linings.

8. State the purpose of a profile frame in relation to 'fixed-in' frames.

9. Describe the procedure for cutting and fixing a timber window board.

10. State the purpose of using frame sealant.

Installing Frames and Linings **Chapter 5**

6

Installing Side Hung Doors

This chapter is intended to provide the reader with an overview of the types of door in use and the procedures used for hanging them. Its contents are assessed in Install Side Hung Doors NVQ unit VR 06.

In this chapter you will cover the following range of topics:

◆ Doors
◆ Hinges
◆ Door and ironmongery schedules
◆ Door hanging

What's required in VR 06?

To successfully complete this unit you will be required to demonstrate your skill and knowledge of the following processes:

◆ Interpreting information
◆ Adopting safe and healthy working practices
◆ Selecting materials, components and equipment
◆ Preparing and hanging side hung doors

You will be required practically to:

◆ Prepare and hang doors:
  ▶ internal and/or external;
  ▶ fire resisting and/or non-fire resisting;
  ▶ single and/or double.
◆ Use hand and portable power tools.
◆ Communicate with other team members.
◆ Undertake calculations concerned with a quantity of materials.

Doors

Doors are moveable barriers, used to cover an opening in a structure. Their main function is to allow access to and from a building and passage between its interior spaces. Other functional requirements include weather protection, fire resistance, sound and thermal insulation, security, privacy, ease of operation and durability. They are often classified by their method of construction and method of operation.

Method of construction

Doors classified by their method of construction are termed matchboarded, panelled, glazed, flush or fire resistant, etc. (Figure 6.1).

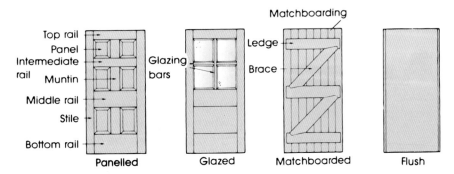

Figure 6.1 *Doors*

did you know?

Framed matchboarded doors constructed with the addition of stiles and rails are used where extra strength is required.

Matchboarded doors – (also termed batten doors) are used mainly externally for gates, sheds and industrial buildings. They are simply constructed from matchboarding, ledges and braces, clench-nailed together. The bottom end of the braces must always point towards the hanging edge of the door to provide the required support. Normally side hung on strap hinges, using tee hinges for side gates and shed doors, or hook and band for larger heavy weight gates and commercial doors.

Panelled doors – have a frame made from solid timber rails and stiles, which are jointed using either dowels or mortise and tenon joints. The frame is either grooved or rebated to receive one or more thin plywood or timber panels. Interior doors are thinner than exterior doors. Normally side hung using butt hinges: 2 for internal doors and 3 for external.

Glazed doors – are used where more light is required. They are similarly made to panelled doors except glass replaces one or more of the plywood or timber panels. Glazing bead is used to secure the glass into its glazing rebates. Glazing bars may be used to divide large glazed areas. As with panelled doors, they are normally side hung using butt hinges, 2 for internal doors and 3 for external.

Flush doors – are made with outer faces of plywood or hardboard. Internal doors are normally lightweight with a hollow core, and solid timber edges termed lippings and blocks, which are used to reinforce hinge and lock positions. New flush doors will have one edge marked 'LOCK' and the other 'HINGE'; these must be followed. External and fire resistant flush doors are much heavier, as most have a solid core of either timber strips or chipboard. A variation on flush doors is to use the same lightweight hollow core but have the faces covered in moulded or embossed facings to give the appearance of a traditional panel door. Internal doors use pressed hardboard facings while external use plastic or metal facings. Again, internal lightweight doors are side hung using 2 butt hinges and 3 for external or heavyweight doors.

Fire resisting doors – are mainly constructed as solid core flush doors. The main function of this type of door is to act as a barrier to a possible fire by providing the same degree of protection as the element in which it is located. They should prevent the passage of smoke, hot gases and flames for a specified period of time. This period of time depends on the relevant statutory regulations and the location of the door. Fire doors are not normally purpose-made, as they must have approved fire resistance certification. It is advantageous to use proven proprietary products. Oversize fire door 'blanks' are available for cutting down to size to suit specific situations.

Fire doors are termed as FD20, FD30 or FD60, which refer to the amount of time in minutes that it can resist the passage of flames and hot gases from the side exposed to a fire through to the un-exposed side (known as its integrity). Intumescent seals may be fitted to the door lipping or frame, which expand when activated by heat in the early stages of a fire to seal the cap between the door and frame, thus helping to prolong the integrity. FD20 and FD30 doors are normally hung using 2 butt hinges and FD60 doors are hung using 3.

Method of operation

A door's method of operation will be determined by its location, construction and desired performance requirements, these include swinging (also termed side hung), sliding and folding as illustrated in Figure 6.2.

> **did you know?**
>
> Details of the door and ironmongery to use in a particular location can be found by reference to the door/ironmongery schedules and the contract specification.

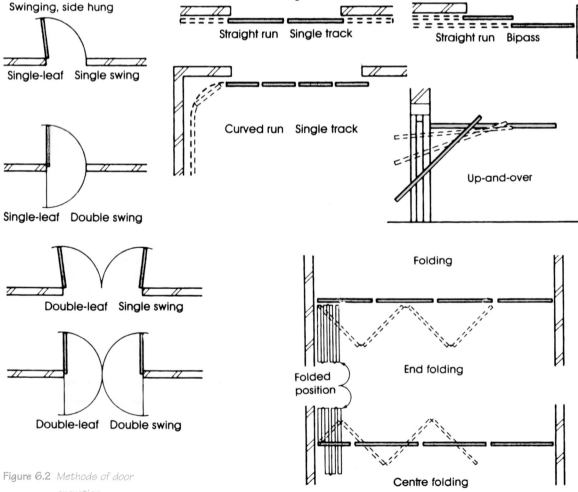

Figure 6.2 *Methods of door operation*

Swinging doors (side hung) – are the most suitable doors for pedestrian use and the most effective for weather protection, fire resistance, sound and thermal insulation. Side hanging on hinges is the most common means of door operation, although pivoted floor springs are more effective where constant use is expected, e.g. shops and office reception areas.

Sliding doors – are mainly used either to economise on space where it is not possible to swing a door, or for large openings which would be difficult to close off with swinging doors.

Folding doors – are a combination of swinging and sliding doors. They can be used as either movable internal partitions to divide up large rooms, or as doors for large warehouses and showroom entrances.

Hinges

Hinges are used to provide the opening and closing action required for doors, windows and cabinet work (Figure 6.3).

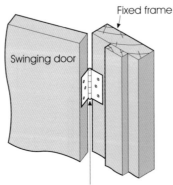

Figure 6.3 *Hinge application*

Hinges

These are available in a variety of materials:

◆ mild steel;
◆ stainless steel;
◆ cast iron;
◆ brass;
◆ plastic.

Pressed mild steel hinges – are the least expensive and are commonly used for internal painted doors. They are also available in different finishes such as electro-brass, bright zinc or chrome plated.

Solid brass and stainless steel hinges – are used for hardwood clear finished doors, both internally and externally. Steel hinges should not be used on hardwood or external doors, because of rusting and subsequent timber staining.

Cast iron hinges – are more suited to painted heavyweight doors. However, take care when fitting as they are brittle and easily broken if hit with a hammer.

did you know?

Plastic and certain brass hinges and other low melting point materials are not suitable for use on fire doors.

Hinge types

Always consult manufacturers' details regarding the suitability of hinges for a particular end use (Figure 6.4).

Butt hinges – are a general-purpose hinge suitable for most applications. They consist of two leaves, joined by a pin that passes through the knuckle formed on the edge of both leaves. As a general rule the leaf with the greatest number of knuckles is fixed to the doorframe.

Flush hinges – can be used for the same range of purposes as butt hinges, on both cabinets and full-size room doors. They are only really suitable for lightweight doors, but they have the advantage of easy fitting, as they do not require 'sinking in'.

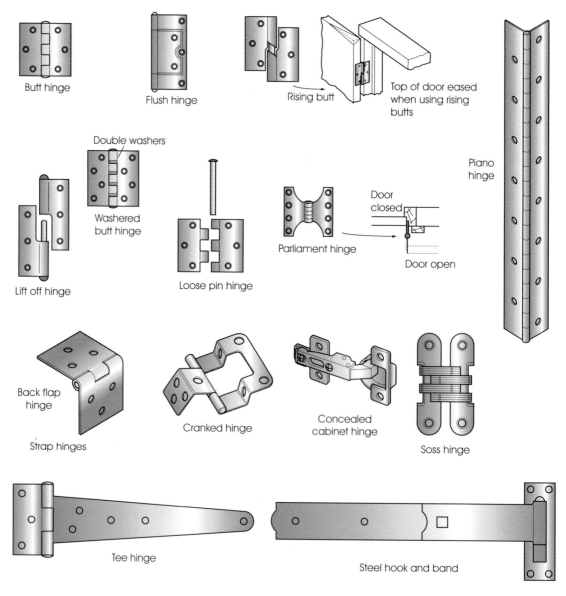

Figure 6.4 *Hinges*

Loose pin butt hinge – enables easy door removal by knocking out the pins. For security reasons they should not be used for outward opening external doors.

Lift-off butt hinges – also enable easy door removal, as the door can be lifted off when in the open position. These are available as right or left-handed pairs. Each pair consists of one long pin hinge, which is fitted as the lower hinge. The upper hinge has a slightly shorter pin, to aid the repositioning of the door.

Washered butt hinges – are used for heavier doors to reduce knuckle wear and prevent squeaking.

Parliament hinges – have wide leaves to extend knuckles and enable doors to fold back against the wall thus clearing deep architraves, etc.

Rising butts – are designed to lift the door as it opens to clear obstructions such as mats and rugs. They also give a door some degree of self-closing action. In order to prevent the top edge of the door fouling the frame as it opens and closes, the top edge must be eased. The hand of the door must be stated when ordering this item (see Chapter 7 for handing details).

Back flap hinges – have extra wide leaves to spread the screw fixing over a large area and are usually used for drop flaps and box lids.

Piano hinges – are made in a continuous 2 m length, which is cut to fit the workpiece. They are used where support is required along the whole length of the hinged joint.

Cranked hinges – are used in cabinetwork for lay-on doors. They enable the door to open through an arc of 180°.

Concealed cabinet hinges – are used for kitchen units and other cabinetwork. Various sizes and types are available. All have a circular boss that fits into a blind hole drilled into the door and a mounting plate that is screwed onto the cabinet standard. Adjustment of the door in all three directions is possible, making it simple to align a row of doors. Spring-loaded versions keep the door closed without the use of a catch.

Soss hinges – are available in a range of sizes for hanging full-size doors and lay-on or inset cabinet doors. They are fitted by drilling and recessing into the door and frame. Their main feature is that they are invisible when the door is closed.

Strap hinges – are surface fixed, and screwed or bolted directly to the door and frame or post.

Tee hinges – are used for ledged and braced doors and gates. Decorative versions are also available.

Hook and band hinges – are suitable for more heavy duty use.

Hinge recessing

The leaves of a hinge can be recessed into the door and frame equally, termed half-and-half (Figure 6.5). Alternatively the hinge can be offset, with the front edge of both leaves recessed into the door, leaving a clean unbroken joint line. This all-in the door method is popular for cabinet work. The method used is dependent on personal preference as all are equally good; check with the specification, foreman or customer.

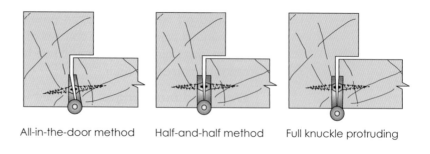

All-in-the-door method   Half-and-half method   Full knuckle protruding

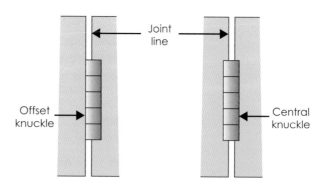

Figure 6.5 *Hinge recessing and knuckle positioning*

did you know?

Cabinet doors often have the leaf recessed up to the centre of the knuckle for a neater appearance. Check the specification and personal preferences.

The position of the knuckle in relation to the face of the door is optional, but is sometimes dependant on the hinge. Full-size room doors are mainly hung with the full knuckle protruding beyond the door face, to give increased clearance.

Recessing procedure

◆ Decide on the hinge position. This will depend on the size and type of door or hinged component to be hung. Small cabinet panel doors and casement sashes normally have the top of the upper hinge in line with the bottom of the top rail. Similarly the bottom of the lower hinge is normally positioned level with the top of the bottom rail.

◆ Mark the hinge position on both the frame and the door or component to be hung (Figure 6.6).

◆ Position the hinge on the edge of the door and mark the top and bottom of the leaf with a sharp pencil.

◆ Repeat the process to mark the leaf position on the frame.

◆ Set a marking gauge to the width of the hinge leaf and score a line on the edge of the door and frame between the two pencil lines. On frames and linings with a stuck rebated stop, a combination square can be set to mark the leaf width with a pencil.

◆ Reset the marking gauge to the thickness of the hinge leaf. Gauge the door face and the frame edge at each hinge position.

◆ Use a chisel held vertically and a mallet to chop the ends of each hinge housing position to the recess.

◆ Use the chisel bevel-side down and a mallet to feather each hinge housing to depth.

did you know?

Remember, when using brass or soft screws to fix ironmongery, insert steel screws first to cut a thread.

◆ With the chisel bevel-side up pare the feathered housing to the gauged depth. Use the chisel at a slight angle so that the bottom of each recess is slightly undercut towards the back.

◆ Screw the hinges to the door or hung component first. Then offer up the door and screw to the frame.

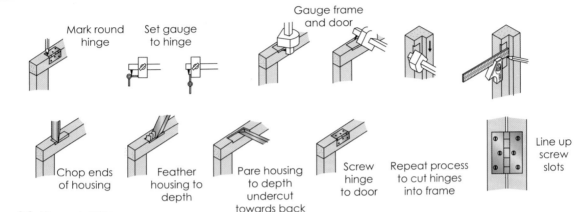

Figure 6.6 *Hinge recessing*

When using brass screws and hinges, it is good practice to screw the hinges on initially with a matching set of steel screws as the softer brass screws are easily damaged and may even snap off. The steel screws will pre-cut a thread into the pilot holes for the brass screws to follow. Candle wax or petroleum jelly may be applied to lubricate the screw thread before insertion. If slot-head screws are being used, these should be lined up vertically to give an enhanced appearance and prevent the build-up of paint or polish in the slots.

Hinge positioning

Lightweight internal doors are normally hung on one pair of 75 mm hinges; glazed, half-hour fire resistant and other heavy doors need one pair of 100 mm hinges. All external doors and one-hour fire resistant doors need one and a half pairs of 100 mm hinges. The standard hinge positions for flush doors are 150 mm down from the top, 225 mm up from the bottom and the third hinge, if required, is positioned centrally to prevent warping, or towards the top for maximum weight capacity (Figure 6.7). On panelled and glazed doors the hinges are often fixed in line with the rails to produce a more balanced look.

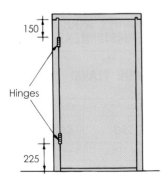

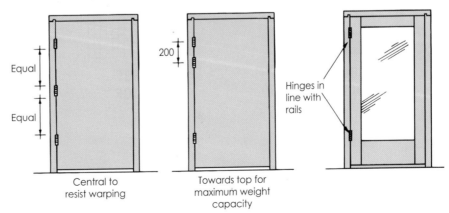

Figure 6.7 *Hinge positioning*

Door and ironmongery schedules

Schedules are used to record repetitive design information. Read these in conjunction with door range drawings and floor plans (Figures 6.8 to 6.10) as they help to identify the type of door, its size, the number required, the door opening in which it fits, the hinges it will swing on and details of any other furniture to be fitted onto it.

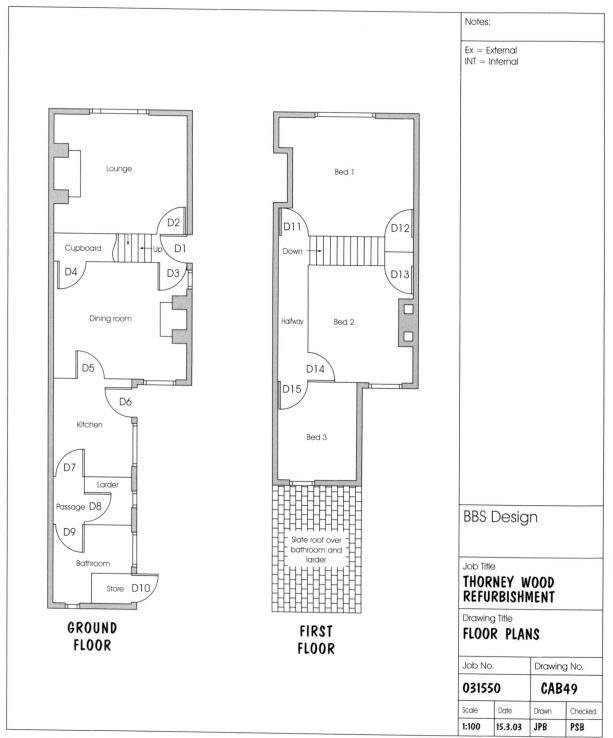

Figure 6.8 *Floor plans*

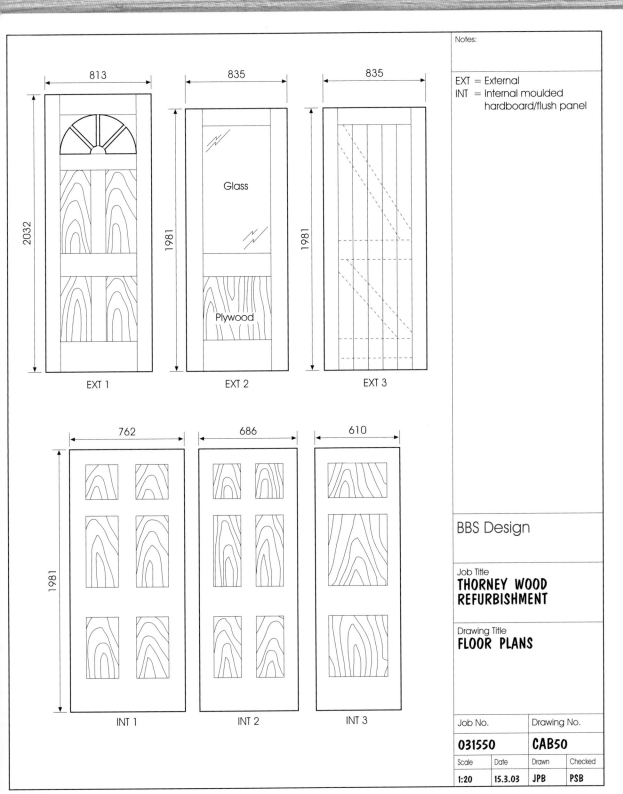

Notes:

EXT = External
INT = Internal moulded
          hardboard/flush panel

813

2032

EXT 1

835

Glass

1981

Plywood

EXT 2

835

1981

EXT 3

762

1981

INT 1

686

INT 2

610

INT 3

BBS Design

Job Title
**THORNEY WOOD
REFURBISHMENT**

Drawing Title
FLOOR PLANS

Job No.	Drawing No.		
031550	**CAB50**		
Scale	Date	Drawn	Checked
1:20	**15.3.03**	**JPB**	**PSB**

Figure 6.9 *Door range drawing*

Chapter 6 Installing Side Hung Doors

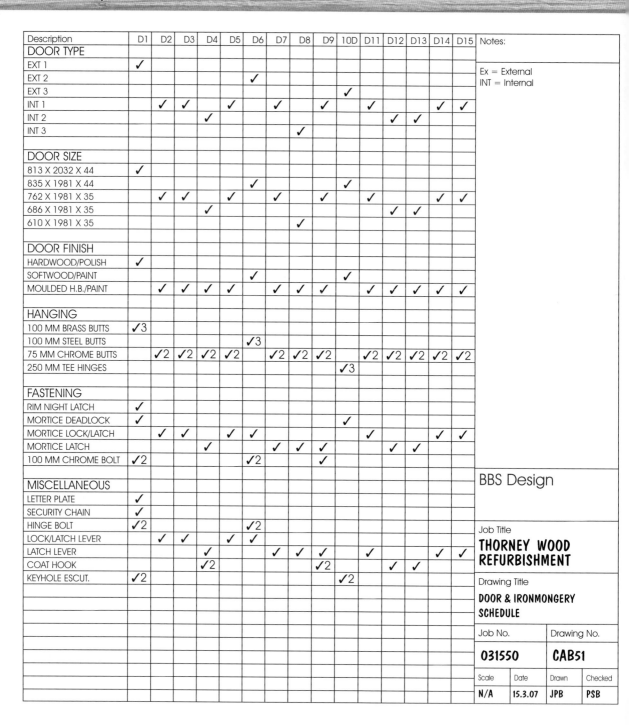

Description	D1	D2	D3	D4	D5	D6	D7	D8	D9	10D	D11	D12	D13	D14	D15
DOOR TYPE															
EXT 1	✓														
EXT 2						✓									
EXT 3										✓					
INT 1		✓	✓		✓		✓		✓		✓			✓	✓
INT 2				✓								✓	✓		
INT 3								✓							
DOOR SIZE															
813 X 2032 X 44	✓														
835 X 1981 X 44						✓				✓					
762 X 1981 X 35		✓	✓		✓		✓		✓		✓			✓	✓
686 X 1981 X 35				✓								✓	✓		
610 X 1981 X 35								✓							
DOOR FINISH															
HARDWOOD/POLISH	✓														
SOFTWOOD/PAINT						✓				✓					
MOULDED H.B./PAINT		✓	✓	✓	✓		✓	✓	✓		✓	✓	✓	✓	✓
HANGING															
100 MM BRASS BUTTS	✓3														
100 MM STEEL BUTTS						✓3									
75 MM CHROME BUTTS		✓2	✓2	✓2	✓2		✓2	✓2	✓2		✓2	✓2	✓2	✓2	✓2
250 MM TEE HINGES										✓3					
FASTENING															
RIM NIGHT LATCH	✓														
MORTICE DEADLOCK	✓								✓						
MORTICE LOCK/LATCH		✓	✓		✓	✓					✓			✓	✓
MORTICE LATCH				✓			✓	✓	✓			✓	✓		
100 MM CHROME BOLT	✓2					✓2			✓						
MISCELLANEOUS															
LETTER PLATE	✓														
SECURITY CHAIN	✓														
HINGE BOLT	✓2					✓2									
LOCK/LATCH LEVER		✓	✓		✓	✓									
LATCH LEVER				✓			✓	✓	✓		✓			✓	✓
COAT HOOK			✓2						✓2			✓	✓		
KEYHOLE ESCUT.	✓2								✓2						

Notes:

Ex = External
INT = Internal

BBS Design

Job Title
THORNEY WOOD REFURBISHMENT

Drawing Title
DOOR & IRONMONGERY SCHEDULE

Job No.	Drawing No.
031550	**CAB51**

Scale	Date	Drawn	Checked
N/A	15.3.07	JPB	PSB

Figure 6.10 *Door and ironmongery schedule*

Details relevant to a particular door opening are indicated in the schedules by a dot, or cross or tick. A figure is also included where more than one item is required. Extracting details from a schedule is called 'taking off'. The following information about the external store door D10 has been taken off the schedules.

> **example**
>
> One softwood painted framed, ledged and braced door type EXT 3 762 mm × 1981 mm × 44 mm, hung on three of 100 mm steel butts.

Take off the following information from the schedules (Figures 6.8 to 6.10): How many type INT 1 painted doors are required?

Produce a list of hinges required for the whole house.

State the size and type of door for opening D5.

Door hanging

Door hanging is normally carried out before skirtings and architraves are fixed. Speed and confidence in door hanging can be achieved by following the procedures below (Figure 6.11).

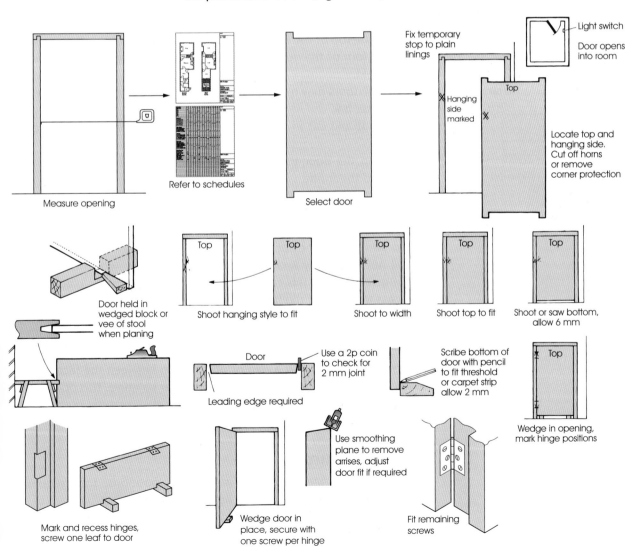

Figure 6.11 *Door hanging procedure*

did you know?

The terms 'shoot in' or 'shoot to fit' mean plane to fit. 'Shoot' comes from the use of a shooting plane to 'true' or straighten long edges.

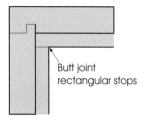

Butt joint rectangular stops

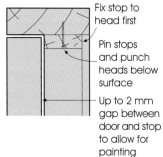

Fix stop to head first

Pin stops and punch heads below surface

Up to 2 mm gap between door and stop to allow for painting

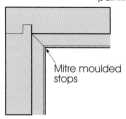

Mitre moulded stops

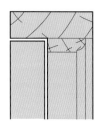

Figure 6.12 *Fixing door stops to plain linings*

- Measure height and width of door opening.
- Refer to door schedule(s) and select the correct door.
- Locate and mark the top and hanging side of the opening and door. For flush and fire-check doors these should have been marked by the manufacturer. When the hanging side is not shown on the drawing, the door should open into the room to provide maximum privacy, but not onto a light switch.
- Cut off the horns (protective extensions on the top and bottom of each stile) on panelled doors. Flush doors will probably have protective pieces of timber or plastic on each corner. These need to be prized off.
- Where plain door linings are used, tack temporary stops either side at mid-height, to stop the door falling through the opening. These should be set back from the edge of the lining, by the door thickness.
- Shoot in (plane to fit) the hanging stile of the door to fit the hanging side of the opening. This should be planed at a slight undercut angle to prevent binding.
- Shoot the door to width. Allow a 2 mm joint all around, between the door and frame or lining. Many woodworkers use a two pence coin to check the 2 mm joint. The closing side will also require planing to a slight angle to allow it to close. This is termed a leading edge.
- Shoot the top of the door to fit the head of the opening. Saw or shoot the bottom of the door to give a 6 mm gap at floor level or to fit the threshold.
- External doors may require rebating along the bottom edge, to fit over the water bar.
- Mark out and cut in the hinges. Screw one leaf of each hinge to the door.
- Offer up the door to the opening and screw the other leaf of each hinge to the frame.
- Adjust the fit as required. Remove all arrises (sharp edges) to soften the corners and to provide a better surface for the paint finish. If the closing edge rubs the frame, the hinges may be proud and so the recesses have to be cut deeper. If the recesses are too deep, the door will not close fully and will tend to spring open, which is known as 'hinge bound'. In this case a thin cardboard strip can be placed in the recess to pack out the hinge.
- Remove temporary stops from plain linings and replace with planted stops pinned in position. Fix the head first, then the sides (Figure 6.12). Rectangular section stops can be simply butted; moulded sections will require mitring. Allow a gap of up to 2 mm between the face of door and the edge of the stop or it will bind after painting. This operation is best done after fitting the lock or latch, as this correctly positions the closing edge. However, it can be done with the help of an assistant.

Weatherboard

Inward opening external doors, in exposed positions, may require a weatherboard (Figure 6.13). These are screwed to the bottom face of the door to throw rainwater clear of the water bar. The ends may be kept clear of the frame or let into it.

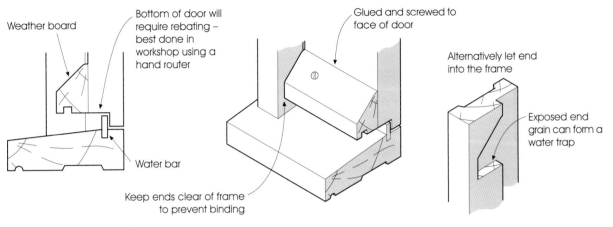

Weather board

Bottom of door will require rebating – best done in workshop using a hand router

Water bar

Keep ends clear of frame to prevent binding

Glued and screwed to face of door

Alternatively let end into the frame

Exposed end grain can form a water trap

Figure 6.13 *Weatherboard for inward opening*

Double doors

The hanging procedure for double doors is similar to that used for single doors, except that only the hanging stiles are planed to fit the frame. The meeting stiles are normally rebated together to provide an overlapping seal rather than a straight joint. The doors should be offered into the opening, keeping a 2 mm gap between the rebates and the hanging stiles marked accordingly. Any surplus timber should be removed evenly from both hanging stiles (Figure 6.14).

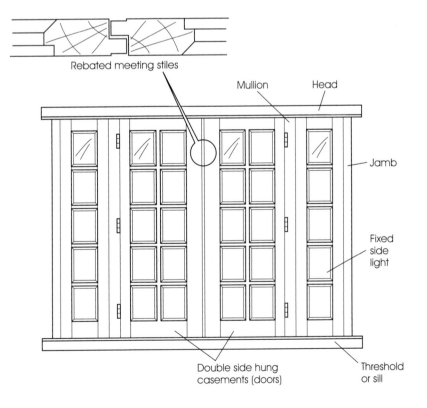

Rebated meeting stiles

Mullion

Head

Jamb

Fixed side light

Double side hung casements (doors)

Threshold or sill

Figure 6.14 *French casements*

Chapter 6 Installing Side Hung Doors

1. State the purpose of door and ironmongery schedules.

2. Produce a sketch to show the typical hinge positions for an external half-glazed door.

3. Explain why, when fitting three hinges to a door, the third hinge can be fitted either centrally between the top and bottom hinge or just below the top hinge.

4. State the purpose of planing a leading edge on a door stile.

5. Define the term 'arris' and state why they should be removed.

6. Statement: Some door edges are fitted with intumescent seals.
   Reason: Intumescent seals act as draft and weatherproofing seals.
   a) Statement true   Reason true
   b) Statement false   Reason false
   c) Statement true   Reason false
   d) Statement false   Reason true

7. Produce a sketch to show the difference between loose pin butt hinges, washed butt hinges and parliament hinges.

8. List a typical sequence of operations for hanging an internal door to a domestic property.

9. In order to prevent sagging in the ledged and matchboarded door shown in Figure 6.15. braces are to be incorporated.

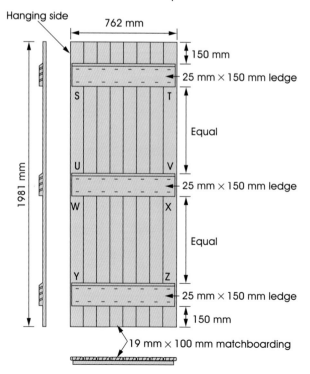

Figure 6.15

Which of the following is the most suitable position for the braces? From:
- **a)** T to U and X to Y
- **b)** S to V and X to Y
- **c)** S to V and W to Z
- **d)** T to U and W to Z

10. Which of the following operations must be carried out when hanging a door using rising butt hinges?
- **a)** Double the leading edge on the closing side
- **b)** Ease the head of the frame on the hanging side
- **c)** Ease the top of the door on the hanging side
- **d)** Increase the clearance on the bottom of the door.

Installing Door Ironmongery

This chapter is intended to provide the reader with an overview of the types of door ironmongery in use and the procedures used for installing them. Its contents are assessed in Install Door Ironmongery NVQ unit VR 07.

In this chapter you will cover the following range of topics:

◆ Door furniture
◆ Ironmongery positioning
◆ Ironmongery schedules
◆ Ironmongery installation procedure

What's required in VR 07?

To successfully complete this unit you will be required to demonstrate your skill and knowledge of the following processes:

◆ Interpreting information
◆ Adopting safe and healthy working practices
◆ Selecting materials, components and equipment
◆ Preparing and fixing ironmongery to side hung doors

You will be required practically to:

◆ Prepare and fix door ironmongery for internal and external doors:
  ► locks;
  ► latches;
  ► cylinder rim latches;
  ► bolts;
  ► letter plates;
  ► door closers.
◆ Use hand and portable power tools.
◆ Communicate with other team members.
◆ Undertake calculations concerned with quantity of materials.

Door furniture

Door ironmongery is also termed 'door furniture' and includes locks, latches, bolts, other security devices, handles and letter or postal plates. Hinges (covered in the previous chapter) are also included in this term. The hand of a door is sometimes required in order to select the correct items of ironmongery.

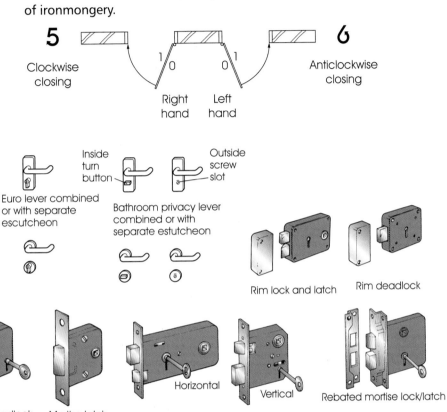

Figure 7.1 *Method for stating door handing*

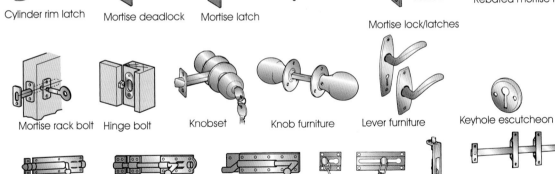

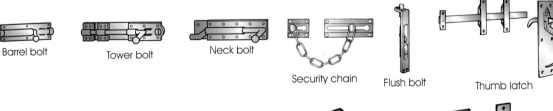

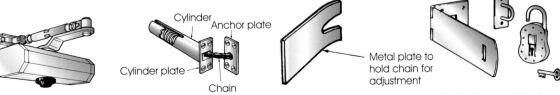

Figure 7.2 *Locks, latches and bolts*

Door handing

Some locks and latches have reversible bolts, enabling either hand to be adopted to suit the situation. View the door from the hinge knuckle side; if the knuckles are on the left the door is left-handed, whereas if the knuckles are on the right, the door is right-handed. Doors may also be defined as either clockwise or anti-clockwise closing when viewed from the knuckle side. When ordering ironmongery simply stating left- or right-hand, clockwise or anti-clockwise can be confusing, as there may be variations between manufacturers and suppliers. The standard way to identify handing is to use the following codes (see Figure 7.1):

◆ 5.0 for clockwise closing doors and indicating ironmongery fixed to the opening face (knuckle side).
◆ 5.1 for clockwise closing doors and indicating ironmongery fixed to the closing face (non-knuckle side).
◆ 6.0 for anti-clockwise closing doors and indicating ironmongery fixed to the opening face.
◆ 6.1 for anti-clockwise closing doors and indicating ironmongery fixed to the closing face.

Locks and latches (see Figure 7.2)

Cylinder rim night latches

These are mainly used for entrance doors to domestic property but, as they are only a latch, they provide little security on their own. When fitted, the door can be opened from the outside with a key and from the inside by turning the handle. Some types have a double locking facility which improves their security. Double locking types are essential on glazed doors, as the inside handle will not turn, even if an intruder breaks a pane to reach inside for it.

Mortise deadlock – This provides a straightforward key-operated locking action and is often used to provide additional security on entrance doors where cylinder rim latches are fitted. They are also used on doors where simple security is required, e.g. storerooms. The more levers a lock has, the better it is. A 5-lever lock is more difficult to pick than a 3-lever.

Mortise latch – used mainly for internal doors that do not require locking. The latch which holds the door in the closed position can be operated from either side of the door by turning the handle.

Mortise lock/latch – two main types are available:

◆ The horizontal is little used nowadays because of its length, which means that it can only be fitted to doors with substantial stiles.
◆ The vertical is more modern and can be fitted to most types of doors. It is often known as a narrow-stile lock/latch.

Both types can be used for a wide range of general-purpose doors in various locations. They are a combination of the mortise deadlock and the mortise latch. A further variation is the Euro pattern mortise lock/latch, which uses a cylinder lock to operate the dead bolt. Bathroom privacy lock/latches are also available, which use a turn button on the inside to operate the dead bolt.

Rebated mortise lock/latch – should be used when fixing a lock/latch in double doors that have rebated stiles. The front end of this lock is cranked to fit the rebate on the stiles. A conversion kit is also available for use with a standard mortise lock/latch.

Rim deadlock

This is a surface fixed lock, traditionally used in the place of a mortise deadlock, but now rarely used.

Rim lock/latch – a surface fixed version of the mortise lock/latch. Mainly used today for garden gates and sheds. Knob furniture is used to operate the latch.

Knobset

A knobset consists of a small mortise latch and a pair of knob handles that can be locked with a key so that it operates as a lock/latch in most situations, both internally and externally. Knobsets are also sold without the lock in the knob for use as a latch only.

Knob furniture – for use with the horizontal rim and mortise lock/latches. It should not be used with vertical types, as hand injuries may result.

Keyhole escutcheon plates

These are used to provide a neat finish to the keyhole of both deadlocks and horizontal mortise lock/latches.

Lever furniture

This is available in a wide range of patterns, for use with the mortise latches and mortise lock/latches, including Euro pattern and bathroom privacy versions. The bathroom privacy version has a turn button on the inside to operate the dead bolt and a turn-screw on the outside of the door to release the dead bolt in emergencies.

Barrel and tower bolts

These secure external gates and doors from the inside. Two bolts are normally used, one near the top of the door and the other near the bottom. Cranked or swan-necked versions are available for use on the inside of outward opening external doors.

Flush bolt – is flush fitting and therefore requires recessing into the timber. It is used for better quality work on the inside of external doors to provide additional security and also on double doors and French windows to bolt one door in the closed position. Two bolts are normally used, one at the top of the door and the other at the bottom.

Mortise rack bolt – a fluted key operated dead bolt, mortised into the edge of a door, at about 150 mm from the top and bottom.

Hinge bolt – helps prevent a door being forced off its hinges, particularly on outward-opening external doors where the hinge knuckle pin is vulnerable.

Security chains

Chains can be fixed on front entrance doors, the slide to the door and the chain to the frame. When the chain is inserted into the slide, the door will only open a limited amount so that the identity of the caller can be checked.

Thumb latch

These are also known as a Norfolk or Suffolk latch. They are used to latch doors and gates in the closed position and can be operated from either side of the door.

Padlock, hasp and staple

These are used on external doors, gates and sheds to provide security. The hasp is fixed to the door and the staple to the frame. Larger types are fixed by bolting through the door rather than screwing, to provide greater protection.

Door closers

These are used to give doors a self-closing action.

Overhead door closers – are mainly used on the medium to heavy weight side-hung doors in offices, shops and industrial premises, to provide a controlled self-closing action. They can also include a hold-open position and temperature-sensitive device to close a held-open door in the event of a fire. The speed of closing can normally be controlled via an adjustment screw, which is located under the closers main cover. Manufacturers provide full installation and maintenance instructions with these items; further adjustment and maintenance information is usually available on their website.

Concealed spring door closers – are used in the hanging stile of light to medium weight and fire-resisting doors to provide a self-closing action. They are suitable for most side hung doors weighing up to 50 kg, except those hung on rising butts, parliament hinges or where limited air displacement conditions exist. The cylinder that contains the spring is mortised into the edge of the door and the cylinder plate is recessed flush. The anchor plate is recessed and fixed to the doorframe. To adjust the closing action:

◆ insert the metal holding plate over the chain against the cylinder plate, with the door in the open position;
◆ un-screw the anchor plate and turn it either clockwise to speed-up the closing action or anti-clockwise to slow it down;
◆ finally re-screw the anchor plate to the frame, then remove the holding plate and check the closing action.

Manufacturers provide full installation and maintenance instructions with these items, plus instructions for further adjustment and maintenance is usually available on their website.

Ironmongery schedules

Details of the ironmongery to be fitted to doors can be found on the door ironmongery schedule (used to record repetitive design information). Ironmongery schedules are sometimes combined with door schedules, as shown in Figure 6.10 Chapter 6, or are provided as a separate document (Figure 7.3).

Installing Door Ironmongery **Chapter 7**

Description	D1	D2	D3	D4	D5	D6	D7	D8	D9	D10			NOTES
Hanging													
Pair 100 mm pressed steel butt hinges			●	●	●.5								
Pair 100 mm brass butt hinges	●.5												
Pair 75 mm pressed steel butt hinges						●	●	●	●.5	●			
Pair 75 mm brass butt hinges		●											
Fastening													
Rim night latch	●												
Mortise deadlock	●												
Mortise lock/latch		●			●				●				
Mortise latch			●	●		●	●	●					
100 mm brass bolts	●²				●²								BBS DESIGN
Miscellaneous													
Brass lock/latch furniture		●			●				●				
Brass latch furniture			●	●		●	●	●					JOB TITLE — PLOT 3 Hilltop Road
Brass letterplate	●												DRAWING TITLE — Ironmongery schedule/doors
Brass knocker	●												JOB NO. DRAWING NO.
Brass coat hook		●²							●²				
Brass escutcheon	●²												
													SCALE DATE DRAWN CHECKED

Figure 7.3 *Door schedules*

Ironmongery positioning

The position of door ironmongery or furniture depends on the type of door construction, the specification and the door manufacturer's instructions (Figure 7.4).

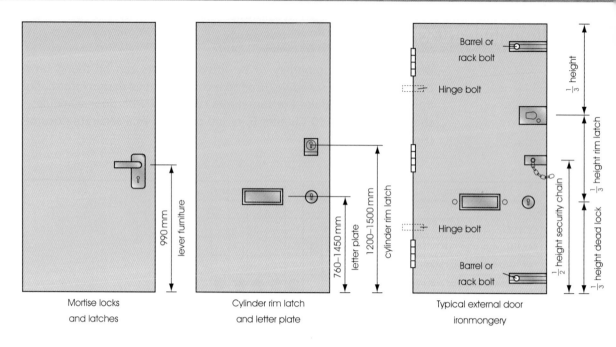

Mortise locks and latches

Cylinder rim latch and letter plate

Typical external door ironmongery

Figure 7.4 *Ironmongery positioning*

did you know?

Always read the job specification carefully as exact furniture positions may be stated.

◆ The standard position for mortise locks and latches is 990 mm from the bottom of the door to the centreline of the lever or knob furniture spindle. However, on a panelled door with a middle rail, locks/latches may be positioned centrally in the rail's width.

◆ Cylinder rim latches are positioned in the door's stile between 1200 mm and 1500 mm from the bottom of the door and the centreline of the cylinder.

◆ Before fitting any locks/latches, the width of the door stile should be measured to ensure the lock/latch length is shorter than the stile's width, otherwise a narrow stile lock may be required.

◆ On external doors with both a cylinder rim latch and a mortise dead lock, the best positions for security is one-third up from the bottom for the dead lock and one-third down from the top for the cylinder rim latch.

◆ Security chains are best positioned near the centre of the door in height.

◆ Hinge bolts should be positioned just below the top hinge and just above the bottom hinge.

◆ Letter plates are normally positioned centrally in a door's width and between 760 mm and 1450 mm from the bottom of the door to the centre line of the plate. Again, on panelled doors letter plates can be positioned centrally in a rail or sometimes vertically in a stile.

Before starting work, read both the door manufacturer's instructions (Figure 7.5) and the ironmongery manufacturer's instructions (Figure 7.6) to ensure the intended position is suitable to receive the item, e.g. the positioning of the lock block on a flush door, and the correct fixing procedure.

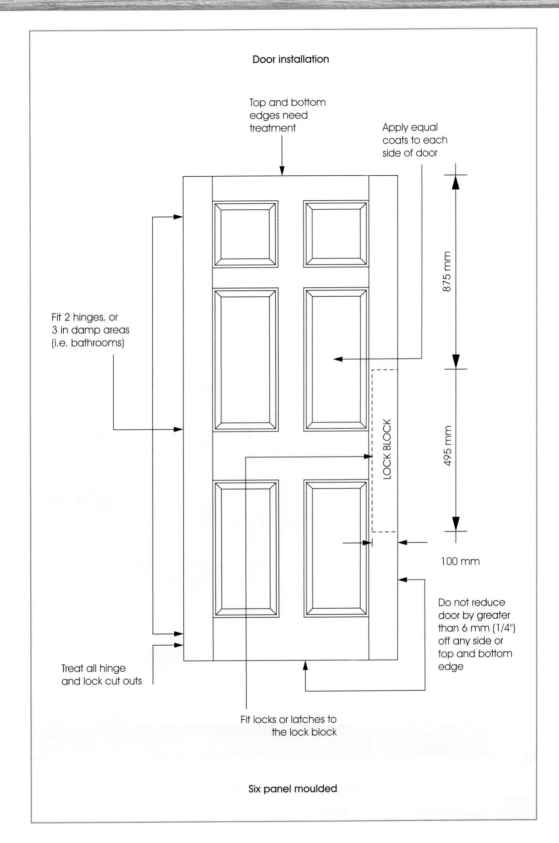

Door installation

Top and bottom edges need treatment

Apply equal coats to each side of door

875 mm

Fit 2 hinges, or 3 in damp areas (i.e. bathrooms)

LOCK BLOCK

495 mm

100 mm

Do not reduce door by greater than 6 mm (1/4") off any side or top and bottom edge

Treat all hinge and lock cut outs

Fit locks or latches to the lock block

Six panel moulded

Figure 7.5 *Typical manufacturer's door installation instructions*

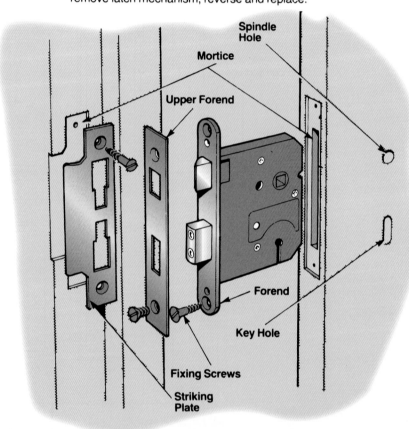

Fixing Instructions

1. At the desired height drill a 15 mm ($^{19}/_{32}$″) diameter spindle hole at a distance of 44 mm (1¾″) from the edge of the door; 57 mm (2¼″) for 3″.
2. At the position of the key hole cut away a section to suit, in line with the spindle hole.
3. Mortice door for lock case and forend.
4. Remove upper forend by removing small lug screws.
5. Fit mortice lock and secure with screws provided.
6. Replace upper forend. Align striking plate with lock and mortice frame to suit.
7. Fix striking plate with two wood screws.

NOTE:
The bolt is reversible. Simply remove lock case, remove latch mechanism, reverse and replace.

Figure 7.6 *Typical manufacturer's fixing instructions*

Ironmongery installation procedure

did you know?

Keyholes should be cut to guide the key. A large round hole would be like putting a pencil in a bucket!

Installation of mortise dead lock, latch or lock/latch

- Wedge the door in the open position (Figure 7.7). Use the lock as a guide to mark the mortise position on the door edge at the pre-marked height.
- Set a marking gauge to half the thickness of the door. Score a centre line down the mortise position to mark for drilling.

◆ Select a drill bit the same diameter as the thickness of the lock body. An oversize bit will weaken the door: use an undersize one and additional paring will be required. A hand brace and bit or power drill and spade bit may be used. Mark the required drilling depth on the bit using a piece of masking tape. Drilling too deep will again weaken the door as well as increase the risk, in panel and glazed doors, of the drill breaking out through the stile.

◆ Drill out the mortise, working from the top; each hole should slightly overlap the one before.

◆ Use a chisel to pare away waste between the holes to form a neat rectangular mortise.

◆ Slide the lock into the mortise and mark around the faceplate. Remove the lock and use a marking gauge to score deeper lines along the grain, as this helps to prevent the fine edge breaking out or splitting when chiselling out the housing.

◆ Use a chisel to form the housing for the faceplate (let-in).

◆ The surface can first be feathered (as when recessing hinges) and finally cleaned.

◆ Hold the lock against the face of the door, with the faceplate flush with the door edge and lined up with the faceplate housing. Mark the centre positions of the spindle and keyhole as required with a bradawl.

◆ Use a 16mm drill bit to drill the spindle hole, working from both sides to avoid breakout, or clamp a waste piece to the back of the door.

◆ Use a 10mm drill bit to drill the keyhole, again working from both sides. Cut the key slot with a pad saw and clean out with a chisel to form a key guide. Alternatively, use a 6mm drill bit to drill out a second hole below the 10mm hole and chisel out waste to form a key guide. Never drill a larger hole for the key as it will not give a guide when inserting the key, making it a hit-and-miss affair.

◆ Insert the lock, check the spindle and keyholes, and align from both sides. Secure the faceplate with screws. Re-check that the key works.

◆ With the dead bolt out, close the door against the frame and mark the bolt and latch positions on the edge of the frame. Square these positions across the face of the frame.

◆ Set the adjustable square from the face of the door to the front edge of the latch or dead bolt. Use it to mark the position on the face of the frame.

◆ Hold the striking plate over the latch or dead bolt position and mark around the striking plate. Gauge vertical lines to prevent breakout when chiselling.

◆ Chisel out to let in the striking plate. Again you may find it easier to feather first before finally cleaning out. The extended lead-in or lip for the latch may require a slightly deeper bevelled housing or recess.

◆ Check for fit and screw the striking plate in place.

◆ Select a chisel slightly smaller than the striking plate bolt-holes. Chop mortises to accommodate both the latch and dead bolt. Some striking plates have boxed bolt-holes. These must be cut beforehand.

◆ Finally, fit the lever furniture, knob furniture or keyhole escutcheon plates as appropriate and check for smooth operation.

◆ When fixing keyhole escutcheon plates, the key should be passed through the plate and into the lock and centralised on the key shaft before screwing.

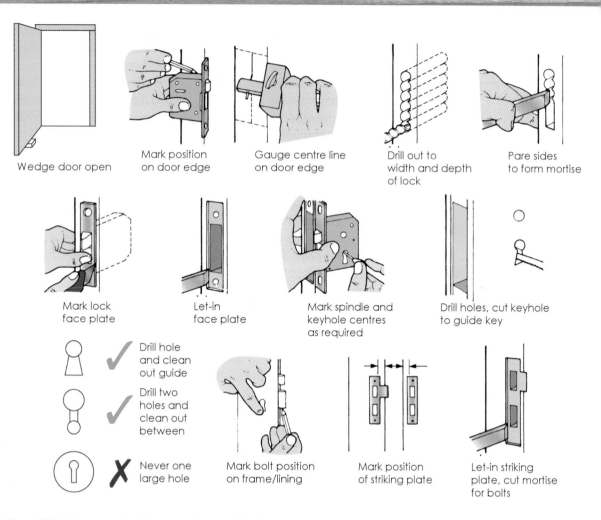

Figure 7.7 *Fitting procedure for mortise locks and latches*

Installation of cylinder rim locks

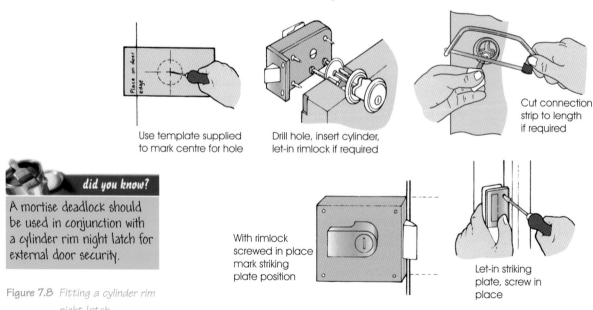

Use template supplied to mark centre for hole

Drill hole, insert cylinder, let-in rimlock if required

Cut connection strip to length if required

With rimlock screwed in place mark striking plate position

Let-in striking plate, screw in place

did you know?

A mortise deadlock should be used in conjunction with a cylinder rim night latch for external door security.

Figure 7.8 *Fitting a cylinder rim night latch*

- Wedge the door in an open position. Use the template supplied with the lock at the pre-marked height to mark the centre of the cylinder hole (Figure 7.8).
- Use a 32 mm auger bit in a brace or a spade bit in a power drill to drill the cylinder hole. Drill from one side until the point just protrudes. Complete the hole from the other side to make a neat hole, avoiding breakout. Alternatively, clamp a block to the door.
- Pass the cylinder through the hole from the outside face and secure it to the mounting plate on the inside with the connecting machine screws.
- Ensure that the cylinder key slot is vertical before fully tightening the screws. For some thinner doors these machine screws may require shortening before use with a hacksaw. If required, secure the mounting plate to the door with woodscrews.
- Check the projection of the flat connection strip. This operates the latch and is designed to be cut to suit the door thickness. If necessary, use a hacksaw to trim the strip so that it projects about 15 mm past the mounting plate.
- Align the arrows on the backplate of the rim lock and the turn-able thimble.
- Place the rim lock case over the mounting-plate, ensuring that the connection strip enters the thimble.
- Mark out and let-in the rim lock lip in the edge of the door if required.
- Secure the rim lock case to the door or mounting plate with wood or machine screws as required. Check both the key and inside handle for smooth operation.
- Close the door and use the rim lock case to mark the position of the keep (striking plate) on the edge of the doorframe.
- Open the door and use the keep to mark the lip recess on the face of the frame. Chisel out a recess to accommodate the keep's lip. Secure the keep to the frame using woodscrews.
- Finally, check from both sides to ensure a smooth operation.

Installation of letter plates

- Wedge the door in the open position. Mark the centreline of the plate on the face of the door (Figure 7.9).
- Position the plate over the centreline and mark around it.
- Measure the size of the opening flap and mark the cut-out on the door. Allow about 2 mm larger than the flap, to ensure ease of operation.
- Mark the position of the holes for the securing bolts.

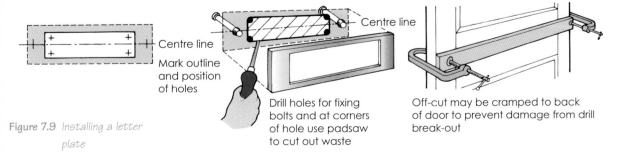

Centre line

Centre line

Mark outline and position of holes

Drill holes for fixing bolts and at corners of hole use padsaw to cut out waste

Off-cut may be cramped to back of door to prevent damage from drill break-out

Figure 7.9 *Installing a letter plate*

- If the door is easily removed, cramp an off-cut of timber to the back of the door, to prevent damage from drill breakout. Alternatively, drill the holes from both sides, with the door in the hanging position.
- Drill holes for the fixing bolts and at each corner of the flap cut-out.
- Use a jig saw or padsaw to saw from hole to hole.
- Neaten up the cut-out using glass paper. Remove the arris from the inside edges.
- Position the letter plate and secure, using the fixing bolts.
- Check the flap for ease of operation and adjust if necessary.

Installation of barrel and tower bolts

- Place the bolt in the required position. Mark one of the screw holes through the backplate with a bradawl or pilot drill.
- Insert a screw, ensuring the bolt is parallel to the edge of the door, and insert a screw at the other end of the bolt (Figure 7.10).

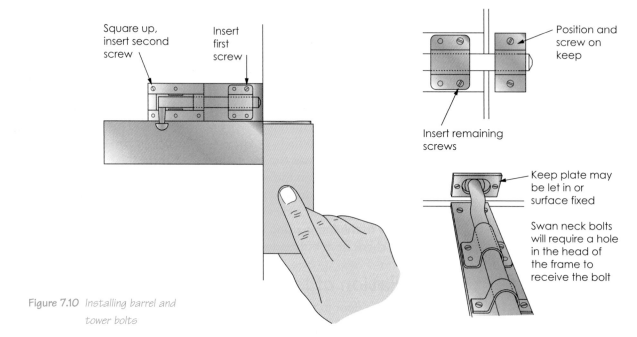

Square up, insert second screw

Insert first screw

Position and screw on keep

Insert remaining screws

Keep plate may be let in or surface fixed

Swan neck bolts will require a hole in the head of the frame to receive the bolt

Figure 7.10 *Installing barrel and tower bolts*

- Move the bolt to the locked position and slide the keep over the bolt.
- Mark the screw holes in the keep and screw in place.
  Check the bolt works smoothly before inserting the remaining screws.

Cranked or swan necked bolts – will require a hole drilling in the head of the doorframe to receive the bolt.

- Position and secure the bolt as before.
- Slide the bolt to the locked position and mark around the bolt.
- Use an auger bit slightly larger than the width of the bolt to drill a hole in the marked position. Ensure the bolt can slide to its full length.
- Check for smooth operation.

Installation of mortise rack bolts

- Use a marking gauge to score a centreline on the edge of the door (Figure 7.11).
- Mark the centre of the bolt-hole using a try square and pencil. Transfer the line onto the face of the door that the bolt will operate from.
- Select an auger or spade bit slightly larger than the bolt barrel. Drill a hole in the edge of the door. Use a piece of tape wrapped around the bit as a depth guide.
- Insert the bolt, and then turn the faceplate so that it is parallel to the door edge. Mark around the faceplate with a pencil.
- Remove the bolt, then gauge the parallel edges. Using a sharp chisel 'let in' the faceplate.
- Place the bolt on the face of the door. Use the bradawl to mark the key position.
- Use a 10 mm bit to drill the keyhole through the face of the door, into the bolt-hole.
- Insert the bolt into the hole then check that the faceplate finishes flush and that the keyhole lines up. Insert the fixing screws.
- Insert the key through the keyhole plate into the bolt. Position the keyhole plate centrally over the key and insert the fixing screws.
- Close the door and turn the key to rack the bolt to its locked position. The point on the bolt will mark the clearance hole centre position on the frame.
- Drill the clearance hole for the bolt in the frame on the marked centre.
- Close the door. Rack the bolt and check for the correct alignment of the bolt clearance hole.
- 'Let in' the keep centrally over the clearance hole and insert the fixing screws.
- Check for smooth operation.

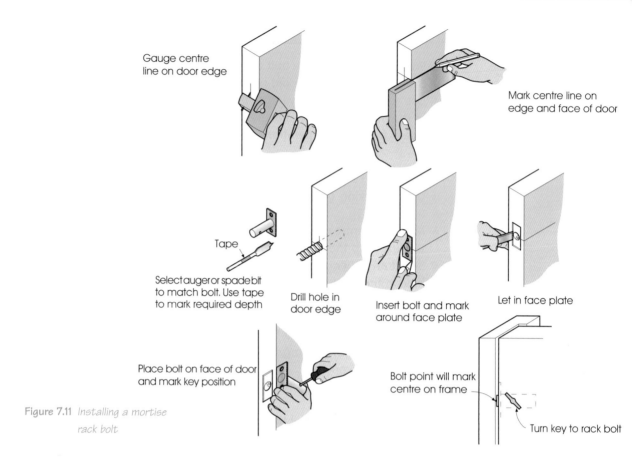

Gauge centre line on door edge

Mark centre line on edge and face of door

Tape

Select auger or spade bit to match bolt. Use tape to mark required depth

Drill hole in door edge

Insert bolt and mark around face plate

Let in face plate

Place bolt on face of door and mark key position

Bolt point will mark centre on frame

Turn key to rack bolt

Figure 7.11 *Installing a mortise rack bolt*

activity

1. You have been asked by your manager to advise a client on the upgrading of security to the external doors of a house.

   At present the inward-opening front entrance door is fitted with a cylinder rim night latch and one barrel bolt. The outward opening kitchen door is fitted with a three-lever mortise lock/latch and one barrel bolt. Both doors are half-glazed softwood with a plywood bottom panel, hung on three hinges, in a softwood rebated frame.

   Write a letter to the client listing your recommendations.

2. The ironmongery schedule shown in Figure 7.3 is for one plot on an estate containing 12 identical houses.

   On a photocopy of the blank order/requisition form, fill in the door ironmongery (excluding hinges) required for the 12 houses. Your order number is PON 1204 and all items are required on-site by 4 weeks from today's date.

   Visit your local ironmonger or look on the Internet for a suitable supplier in order to obtain the costs of the required items. Add this information to the form.

BBS CONSTRUNCTION
ORDER/REQUISITION

Registered office

No. _____

Date _____

To _____

From _____

Address

Site address

_____

_____

Please supply or order for delivery to the above site the following:

Description	Quantity	Rate		Date required by

Site manager/foreman _____

Note Please advise site within 24 hours of request if order cannot be fulfilled by the date required

Figure 7.12 *BBS order form*

measuring up

Chapter 7 | Installing Door Ironmongery

1. State why manufacturer's instructions should be followed when installing ironmongery.

2. Explain the reason for cutting a keyhole to the shape of the key rather than a larger diameter, round hole.

3. Produce a sketch to show the typical positions of a cylinder rim night latch, mortise dead lock and mortise bolts to a front entrance door.

4. Which of the following is the reason for boring a spindle hole from both sides of the door?
    a) the drill is too short
    b) for greater accuracy
    c) to prevent breakout
    d) easy to mark out

5. Explain the difference between a standard pattern and Euro pattern lock/latch.

6. State the reason why the use of double-locking cylinder rim night latches is considered essential for glazed entrance doors.

7. State why the stile of a door should be measured before ordering any ironmongery.

8. State the purpose of an ironmongery schedule.

9. Which of the following items of door ironmongery is illustrated in Figure 7.13?
    a) Vertical mortise lock/latch
    b) Horizontal mortise lock/latch
    c) Euro pattern mortise lock/latch
    d) Vertical rim lock/latch.

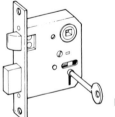

Figure 7.13

10. State the hand of door illustrated in Figure 7.14.

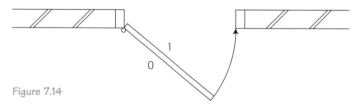

Figure 7.14

Installing Internal Mouldings

This chapter is intended to provide the reader with an overview of installing internal mouldings. Its contents are assessed in Install Internal Mouldings NVQ unit VR 08.

In this chapter you will cover the following range of topics:

- Mouldings and trim
- Cutting and fixing trim
- Installing skirting
- Installing dado and picture rails
- Estimating materials

What's required in VR 08?

To successfully complete this unit you will be required to demonstrate your skill and knowledge of the following processes:

- Interpreting information
- Adopting safe and healthy working practices
- Selecting materials, components and equipment
- Preparing and fixing internal mouldings.

You will be required practically to:

- Prepare and fix:
  - architraves;
  - skirtings;
  - dado rails;
  - picture rails.
- Mitre and scribe mouldings.
- Scribe mouldings to irregular surfaces.
- Return mouldings across the width and thickness.
- Use hand and portable power tools.
- Communicate with other team members.
- Use access equipment.
- Work at height.

Chapter 8 Installing Internal Mouldings

Mouldings and trim

Mouldings – are the ornamental contours or shapes applied to angles or features of building material for decorative purposes. They are variously named according to their profile (Figure 8.1).

Trim – is the collective term for timber moulded sections. They may be termed as either vertical or horizontal and are used to cover the joint between adjacent surfaces, such as wall and floor/ceiling or the joint between plaster and timberwork. In addition, they can provide a decorative feature, and may also serve to protect the wall surface from knocks and scrapes (Figure 8.2).

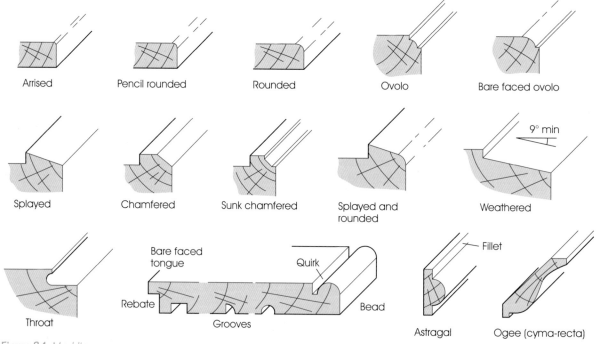

Figure 8.1 *Mouldings*

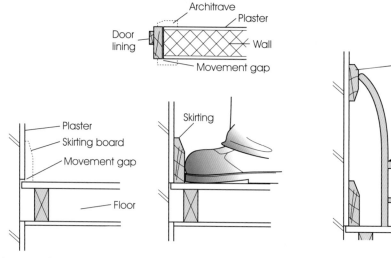

Figure 8.2 *Covering of gaps and protecting plasterwork*

The common types and location of timber trim are illustrated in Figure 8.3.

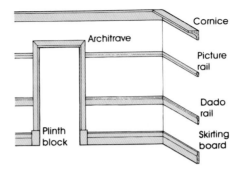

Figure 8.3 *Types of trim*

Architrave – The decorative trim that is placed internally around door- and window-openings to mask the joint between the wall and timber and to conceal any subsequent shrinkage and expansion.

Skirting – The horizontal trim, often a timber board, that is fixed around the base of a wall to mask the joint between the wall and floor. It also protects the plaster surface from knocks at floor level.

Dado rail – A moulding applied to the lower part of interior walls at about waist height, or approximately 1 m from the floor. It is also known as a chair rail as it coincides with a tall chair-back height to protect the plaster.

Picture rail – A moulding applied to the upper part of interior walls between 1.8 m and 2.1 m from the floor. Special clips may be hooked over the rail for hanging picture frames.

Cornice – The moulding used internally at the wall/ceiling junction. It is normally formed from plaster, rarely from timber.

Cutting and fixing trim

Solid timber, either softwood or hardwood, is used for the majority of mouldings. However, MDF (medium density fibreboard) and low density foamed-core plastics are used to a limited extent for moulding production. They are normally ready to install, machined to a range of standard profiles (shapes) by the manufacturer or supplier (Figure 8.4).

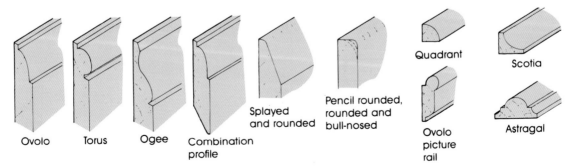

Ovolo   Torus   Ogee   Combination profile   Splayed and rounded   Pencil rounded, rounded and bull-nosed   Quadrant   Scotia   Ovolo picture rail   Astragal

Figure 8.4 *Standard trim sections*

Skirting is often mass-produced, using a combination profile, for example, with an ovolo mould on one face and edge and a splayed and rounded mould on the other. This allows it to be used for either purpose and also reduces the timber merchant's stock range. In addition, the moulding profile provides an undercut edge so that it fits snugly to the floor/wall junction.

The following cutting and fixing details are generally suitable for all three of the above materials. Consult the manufacturer's instructions prior to fixing other proprietary mouldings/trim.

Installing architraves

Architraves are normally cut and fixed after hanging the doors (Figure 8.5). A set of architraves for one side of a room door, consists of a horizontal head and two vertical jambs or legs.

Jointing architraves

A 6mm to 9mm margin is normally left between the frame or lining edge and the architrave (Figure 8.6). This margin neatens the opening; an unsightly joint line would result if architraves were kept flush with the edge of the opening.

The return corners of a set of architraves are mitred (Figure 8.7). For right-angled returns (90°) the mitre is 45° (half the total angle) and cut using either a mitre box/block or mitre frame saw (Figure 8.8). A mitre chopsaw speeds up work when a large number of mitres have to be cut.

Mitres for corners other than right angles should be half the angle of intersection (Figure 8.9). They can be practically found by marking the outline of the intersecting trim on the frame/lining or wall, and joining the inside and outside corners to give the mitre line. Moulding can be marked directly from this; alternatively an adjustable bevel can be used.

The head is normally marked, cut and temporarily fixed in position first. The jambs can then be marked, cut, eased if required and subsequently fixed. Where the corner is not square or you have been less than accurate in cutting the mitre, it will require easing, either with a block plane or by running a tenon saw through the mitre.

Fixing architraves

Fixing is normally direct to the door frame/lining at between 200mm and 300mm centres, using typically 38mm or 50mm long oval or lost-head

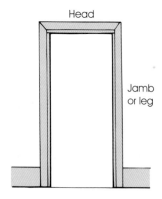

Figure 8.5 Architraves

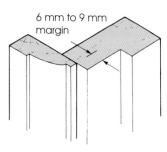

6 mm to 9 mm margin

Figure 8.6 Margin to architraves

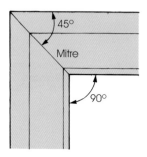

45°

Mitre

90°

Figure 8.7 Mitres to architraves

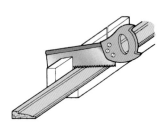

Figure 8.8 Cutting mitres

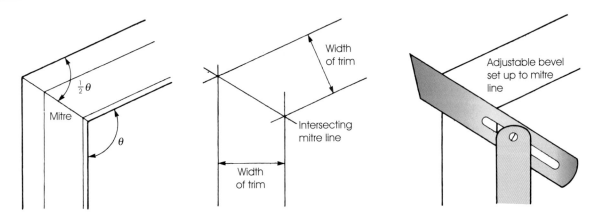

Figure 8.9 *Determining mitre for corners other than right angles*

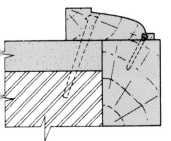

Figure 8.10 *Normal method of fixing architraves*

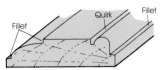

Figure 8.11 *Use fillets or quirks for concealment*

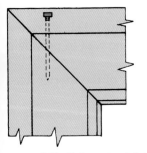

Figure 8.12 *Nailing mitre joints at corners*

safety tip

Do not forget eye protection when driving masonry nails.

nails. These should be positioned in the fillets or quirks (flat surface or groove in moulding) and punched in (Figures 8.10 and 8.11).

Mitres should be nailed through their top edge to reinforce the joint and to ensure both faces are kept flush (Figure 8.12). Use 38-mm oval or lost-head nails. In addition, architraves, especially very wide ones, are often fixed back to the wall surface using either cut or masonry nails.

Architraves to door openings

◆ Mark a margin line around the door opening (Figure 8.13). Check-lines along the width of the architrave can be marked on the wall at the top of the opening.

◆ Starting with the headpiece, cut the mitre on one end, hold the piece in position and mark the second mitre.

◆ Cut the second mitre and temporarily fix in position with two nails. Leave the nails protruding in case later adjustment is necessary.

◆ Hold the first jamb in position and mark the mitre. Cut the mitre, fit with a block plane if required and temporarily nail in position.

◆ Repeat the last stage with the second jamb.

◆ Complete nailing at 200 mm to 300 mm intervals, including nailing across the mitres. Punch all nails below the surface in preparation for filling later by the painter.

◆ Use a glass paper block to remove the arris around the frame, architraves and mitres.

Plinth blocks – were traditionally fixed at the base of an architrave to take the knocks and abrasions at floor level (Figure 8.14).

In current practice, plinth blocks will rarely be used, except for restoration work and new, high-quality traditional-style work. They are also sometimes used to overcome joining problems, which occur when skirtings are thicker than the architrave.

Architraves may be butt jointed to the plinth block, but traditionally they were pre-joined together using a bare-faced tenon and screws.

Corner blocks – are sometimes used at the return corners of a set of architraves in place of mitres (Figure 8.15). These may be simply pinned in place and the architrave butt jointed to them. Alternately, they may be pre-fixed to the head architrave using dowels or a biscuit. To mask the effect of subsequent block shrinkage the head and leg architraves may be housed into them.

Mark margin

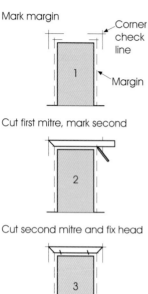

Cut first mitre, mark second

Cut second mitre and fix head

Mark mitre to first jamb

Cut mitre, fix first jamb

Mark mitre to second jamb

Cut mitre, fix second jamb

Punch in nails and remove arrises

Figure 8.13 *Marking and fixing*

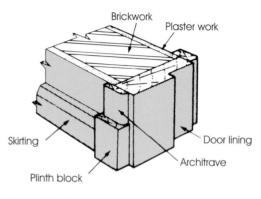

Figure 8.14 *Plinth block*

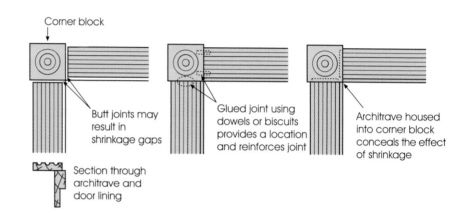

Figure 8.15 *Use of corner blocks*

Architraves to four-sided openings

In addition to the normally three-sided use of architraves for doors, they are also fixed around all four sides of wall serving hatches, ceiling hatches and even wall surfaces without an opening to create a decorative panel effect.

♦ Mark a margin line all around the opening (Figure 8.16). Mark the width of the architrave from the margin line on the wall to provide a check line.
♦ Starting with the top piece for wall hatches or one of the shorter sides of a ceiling hatch.
♦ Cut a mitre on one end of piece (1). Hold the piece in position and mark a second mitre.
♦ Cut the second mitre and temporarily fix in position with two nails. Leave the nail heads protruding in case later adjustment is required.
♦ Cut and fit, if required a mitre on one end of piece (2). Hold the piece in position and mark a second mitre. Cut the mitre and temporarily fix in postion.
♦ Cut and fit, if required, a mitre on one end of the piece (3). Hold the piece in position and mark a second mitre. Cut the second mitre and place the piece aside.

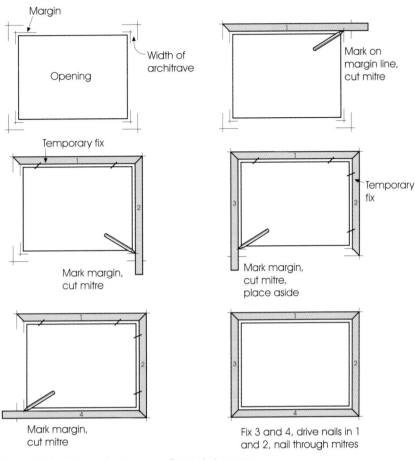

Figure 8.16 *Fixing architraves to a four-sided opening*

Figure 8.17 *Scribing architraves*

- Cut and fit, if required, a mitre on one end of the piece (4). Hold the piece in position and mark a second mitre. Cut the mitre.
- Hold pieces (3) and (4) in position to check the second mitre. Fit if required. Fix both pieces in position.
- Complete nailing at 200 mm to 300 mm centres, including nailing across the mitres. Punch all the nails below the surface, in preparation for filling later by the painter.
- Use a glass paper block to remove arris around the frame, architrave and mitres.

Scribing architraves

Architraves should be scribed (one member cut to fit over the contour of another) to fit the wall surface, where frames/linings abut a wall at right angles.

- Temporarily fix the architrave jamb in position, keeping the overhang the same all the way down. Set a compass to the required margin, plus the overhang, or alternatively use a piece of timber this size as a gauge block (Figure 8.17).
- Mark the line to be cut with the compass or gauge slip.
- Saw along the marked line by slightly undercutting the edge (making it less than 90°) and it will fit snugly to the wall contour (Figure 8.18).

Figure 8.18 *Fitting architrave snugly to wall surface*

Quadrant and Scotia moulds

A quadrant mould or Scotia mould is often used to cover the joint to provide a neat finish to the reveal of external door frames (Figure 8.19). These moulds may also be used in place of an architrave jamb where the frame/lining joins to a wall at right angles. The sharp arris on the back of the mould is best pared off with a chisel, to enable the mould to sit snugly into the plaster/timber intersection.

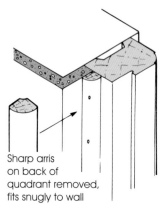

Sharp arris on back of quadrant removed, fits snugly to wall

Figure 8.19 *Use of quadrants as an alternative*

Installing skirting

Skirting is normally cut and fixed directly after cutting and fixing the architraves.

Jointing skirtings

Corners

Internal corners of 90° and less – these right angle and acute angle corners (Figure 8.20) are scribed, with one piece cut to fit over the other (Figure 8.21).

Scribes can be formed in one of two ways:

◆ Mitre and scribe – Fix one piece and cut an internal mitre on the other piece to bring out the profile (Figure 8.22). Cut the profile square on the mitre line to remove waste. Use a coping saw for the curve.
◆ Compass scribe – Fix one piece, place the other piece in position (Figure 8.23). Scribe with a compass. Cut square on the scribed line to remove waste.

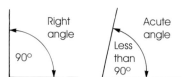

Right angle
90°

Acute angle
Less than 90°

Figure 8.20 *Internal corner 90° or less*

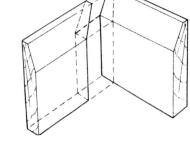

Figure 8.21 *Scribing internal corners*

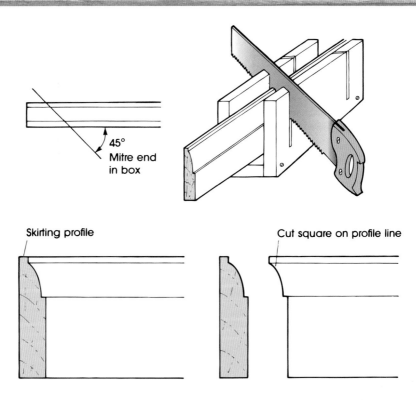

Figure 8.22 *Cutting an internal scribe (mitre and scribe)*

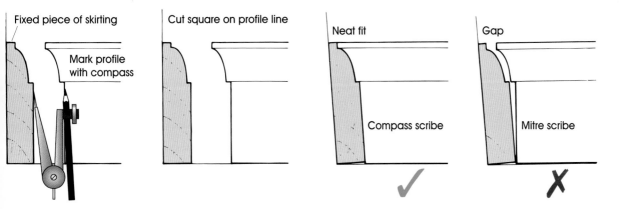

Figure 8.23 *Compass marking and cutting a scribe*

Scribing is the preferred method, especially where the walls are slightly out of plumb. The mitred scribe will have a gap, but the compass scribe will fit the profile neatly.

Mitres are not normally used for internal corners of skirting. Wall corners are rarely perfectly square, making the fitting difficult. In addition, mitres open up as a result of shrinkage, forming a much larger gap than scribes.

However, internal corners on bull-nosed or pencil-rounded skirtings may be cut with a false mitre or partial mitre on the top rounded edge and the remaining flat surface scribed to fit (Figure 8.24).

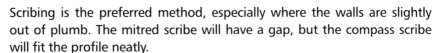

Figure 8.24 *False mitre and scribe*

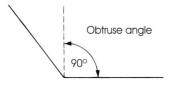

Figure 8.25 *Internal corner over 90°*

Internal corners over 90°

Internal corners over 90° (called obtuse angles, Figure 8.25) are best jointed with a mitre.

Mitring external corners

These return the moulding profile at a corner rather than a butt joint (Figure 8.26), which would otherwise show unsightly end grain (Figure 8.27). Mitres for 90° external corners can be cut in a mitre box or with the aid of a frame saw or mitre chop saw.

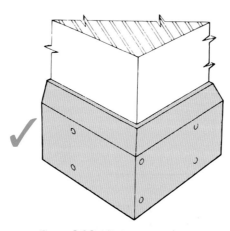

Figure 8.26 *Mitring external corners*

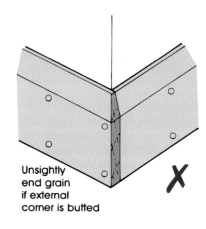

Unsightly end grain if external corner is butted

Figure 8.27 *Butting of external corners not recommended*

Mitres for both internal and external corners over 90° can be marked out using the following method and then cut freehand:

◆ Use a piece of skirting to mark the line and the width of the skirting on the floor either side of the mitre (Figure 8.28).

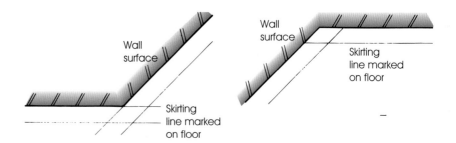

Figure 8.28 *Line of skirting for corner over 90°*

◆ Place the length of the skirting in position.
◆ Mark the position of the plaster arris on the top edge of the skirting. For internal corners this will be the actual back edge of the skirting (Figures 8.29 and 8.30).

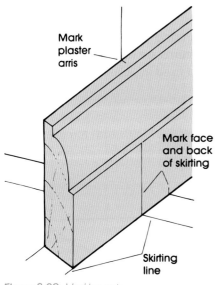

Figure 8.29 *Marking out external corners over 90°*

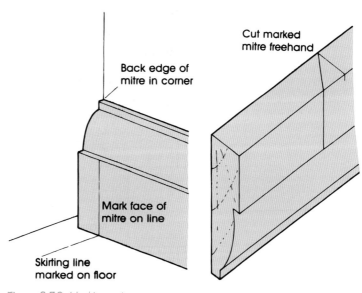

Figure 8.30 *Marking out internal corners over 90°*

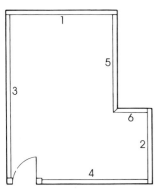

Figure 8.31 *Order of fixing (trapped ends first)*

- Mark the outer section on the front face of the skirting.
- Use a try square to mark a line across the face and back surface of the skirting.
- Cut the mitre freehand or with the aid of a frame saw or chopsaw set to the required angle.
- Repeat the marking out and cutting procedure for other piece.
- Fix the skirting to the wall; external corners should be nailed through the mitre.

Long lengths are fixed first (Figure 8.31), starting with those which have two trapped ends (both ends between walls). Marking and jointing internal corners is much easier when one end is free.

Where the second piece to be fixed also has two trapped ends, a piece slightly longer than the actual length required, by say 50 mm, can be angled across the room or allowed to run through the door opening for scribing the internal joint (Figure 8.32 and 8.33).

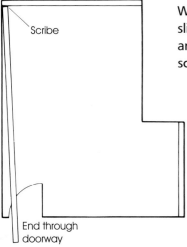

Figure 8.32 *Extend through doorway to permit scribing of joint*

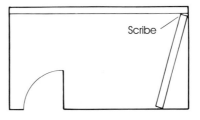

Figure 8.33 *Angle and scribe when second piece has both ends trapped*

After scribing and cutting to length it can be fixed in position.

Very short lengths of skirting returned around projections may be fixed before the main lengths. The two short returns are mitred at their external ends, cut square at their internal ends and fixed in position (Figure 8.34).

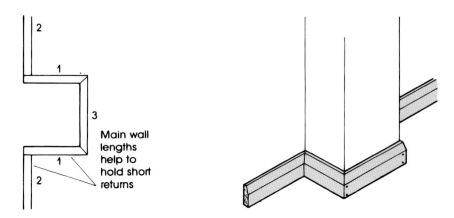

Figure 8.34 *Small pieces may be fitted first*

Main wall lengths are then scribed and fixed in position. These help to hold the short returns. Finally, the front piece is cut and fixed in position by nailing through the mitres.

Heading joints can be used where sufficiently long lengths of skirting are not available. Mitres are preferred to butts, because the two surfaces are held flush together by nailing through the mitre (Figure 8.35). In addition, mitres mask any gap appearing as a result of shrinkage.

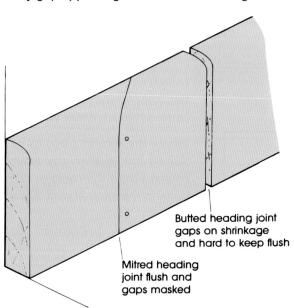

Figure 8.35 *Mitres are preferred for heading joints*

Curved surfaces

Where skirtings and other trim are to be fixed around a curved surface, the back face will almost certainly require kerfing. This involves putting saw cuts in the back face at regular intervals to effectively reduce the thickness of the trim. The kerfs, cut with a tenon saw, should be spaced between 25 mm and 50 mm apart. The tighter the curve, the closer together the kerfs should be. The depth of the kerfs must be kept the same. They should extend through the section to the maximum extent, but kept just back from the face and top edge (Figures 8.36 and 8.37).

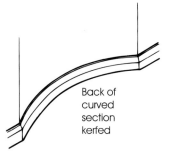

Figure 8.36 *Fixing to a curved surface*

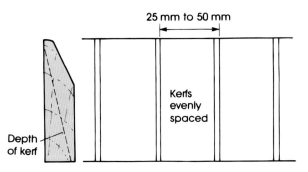

Figure 8.37 *Kerfing to the back of a skirting board*

Care is required when fixing trim back to the wall to ensure that it is bent gradually and evenly. It is at this time that accuracy in cutting the kerfs evenly spaced and to a constant depth is rewarded. Any overcutting and the trim is likely to snap at that point.

Mitres at the ends of curved sections can be marked out and cut using the same methods as described for other obtuse angles.

Returned stop mouldings

Skirtings and other mouldings may occasionally be required to stop part way along a wall, rather than finish into a corner or another moulding such as the skirting to a landing. In these circumstances the profile should be returned to the wall or floor.

Returned to wall – this can be achieved by mitring the end and inserting a short mitred return piece, producing a true mitre for good quality work. Alternatively, the return profile can be cut across the end grain of the main piece; however, end grain will remain exposed (Figure 8.38).

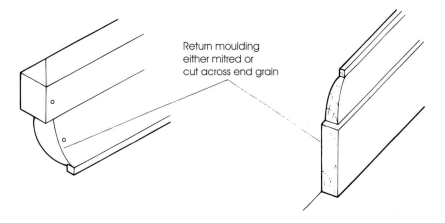

Figure 8.38 *Return the profile of moulding that stops part way along a wall*

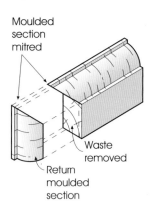

Figure 8.39 *Stopped moulding returned to floor*

Returned to floor – this involves mitring the moulded section (Figure 8.39), cutting out the waste and inserting a return moulded section (cut from an off-cut) down the end of the board.

Fixing skirtings

Skirting can be fixed back to walls with the aid of:

- grounds;
- timber twisted plugs;
- direct to the plaster/wall surface.

Grounds – are timber battens which are fixed to the wall surface using either cut nails in mortar joints or masonry nails. One ground is required for skirtings up to 100 mm in depth. Deeper skirtings require either the addition of vertical soldier grounds at 400 mm to 600 mm centres or an extra horizontal ground (Figure 8.40). The top ground should be fixed about 10 mm below the top edge of the skirting.

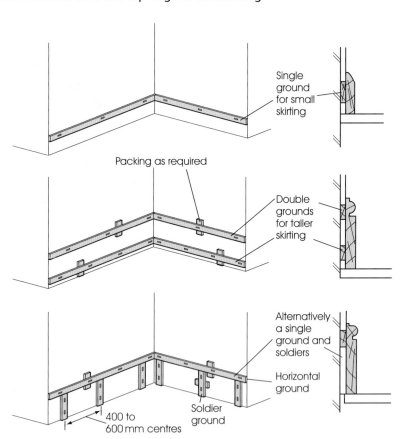

Figure 8.40 *Timber grounds for fixing skirting*

Packing pieces may be required behind the grounds to provide a true surface on which to fix the skirting. Check the line of ground with either a straight-edge or string line. Skirtings can be fixed back to grounds using, typically, 38 mm or 50 mm oval or lost-head nails.

Twisted timber plugs – are rarely used today. They are shaped to tighten when driven into the raked-out vertical brickwork joints at approximately 600 mm apart (Figure 8.41). When all the plugs have been fixed, they

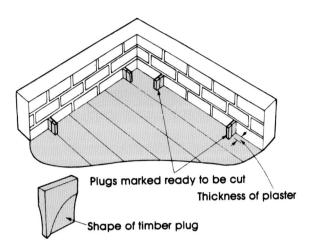

Plugs marked ready to be cut

Thickness of plaster

Shape of timber plug

Figure 8.41 *Fixing skirting to timber plugs*

should be marked with a straight-edge or string line and cut off to provide a true line. An allowance should be made for the thickness of the plaster.

Skirtings are fixed back into the end grain of the plugs using, typically, 50 mm cut nails. These hold better in the end grain than would oval or lost-head nails.

Direct to the wall – skirtings are fixed back to the wall after plastering, using typically either 50 mm cut nails or 50 mm masonry nails, depending on the hardness of the wall. Oval or lost-head nails may be used to fix skirtings to timber studwork partitions.

Fixings should be spaced at 400 mm to 600 mm centres (Figure 8.42). These should be double nailed near the top and bottom edge of the skirting or they can be staggered between the top and bottom edge. Remember that all nails should be punched below the surface.

Hardwood skirtings for very high quality work may be screwed in position. These should either be counter-bored and filled with cross-grained pellets on completion, or brass screws and cups should be used (Figure 8.43).

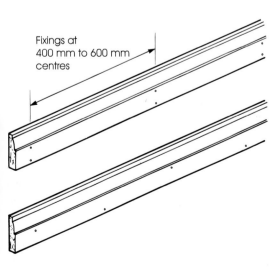

Fixings at 400 mm to 600 mm centres

Figure 8.42 *Spacing of fixings*

Cup and screw

Counter-bored and pelleted

Figure 8.43 *Screws sometimes used to fix hardwood skirting*

As a modern alternative to nails and screws, skirting and other trim fixed to wall surfaces can be bonded using a gun-applied gap-filling, 'no nails' adhesive. One or two continuous 6-mm beads should be applied to the back of the trim before positioning and pressing in place. Strutting or temporary nails (Figure 8.44) may be necessary to hold the trim in place whilst the adhesive cures. These should be left overnight before removal.

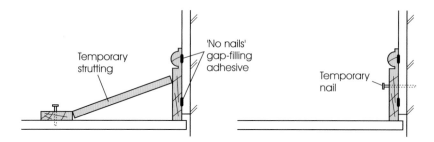

Figure 8.44 *Use of gap-filling adhesive to fix skirting*

Check for hidden/buried services

Prior to fixing mouldings across any wall, a check should be made to see if any services are hidden below the wall surface. Wires to power points normally run vertically up from the floor. Wires to light switches normally run vertically down from the ceiling. Therefore, keep clear of these areas when fixing. Buried pipes in walls are harder to spot.

Vertical pipes may just be seen at floor level; outlet points may also be visible (Figure 8.45). Assume both of these run the full height of the wall, both up and down. Therefore, again keep clear of these areas when fixing. When in doubt an electronic device can be used to scan the wall surface prior to fixing. This gives off a loud noise when passing over buried pipes and electric cables.

Holding down and scribing skirtings

When fixing narrow skirtings, say 75 mm to 100 mm in depth, they may be kept tight down against a fairly level floor surface with the aid of a kneeler (Figure 8.46). This is a short piece of board placed on the top edge of the skirting and held firmly by kneeling on it.

Prior to fixing, check for hidden services: fixing into electric cables and gas or water pipes is potentially dangerous, and expensive to repair.

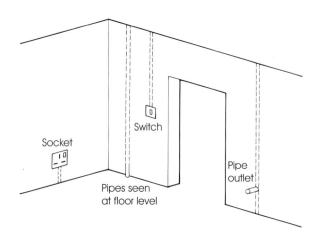

Figure 8.45 *Locating hidden services*

Figure 8.46 *Keeping skirting tight to floor surface*

Deeper skirtings and/or uneven floor surfaces may require scribing to close the gaps before fixing. This is carried out after jointing but prior to fixing:

- ◆ Place cut length of skirting in position.
- ◆ Use gauge slip or compass set to widest gap to mark on the skirting a line parallel to the uneven surface (Figure 8.47).
- ◆ Trim the skirting to line using either a handsaw or a plane.
- ◆ Undercut the back edge to ensure the front edge snugly fits the floor contour (Figure 8.48).

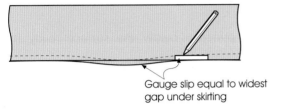

Gauge slip equal to widest gap under skirting

Figure 8.47 Scribing skirting to an uneven floor surface

Edge undercut for snug fit to floor

Figure 8.48 Undercutting bottom edge of skirting

Installing dado and picture rails

Being horizontal mouldings, these may be cut and fixed using similar methods to those used for skirtings.

Start by marking a level line in the required position around the walls (Figure 8.49). Use a straight-edge and spirit level or a water level and chalk line. The required position may be related to a datum line where established.

When working single-handed, temporary nails can be used at intervals to provide support prior to fixing (Figure 8.50).

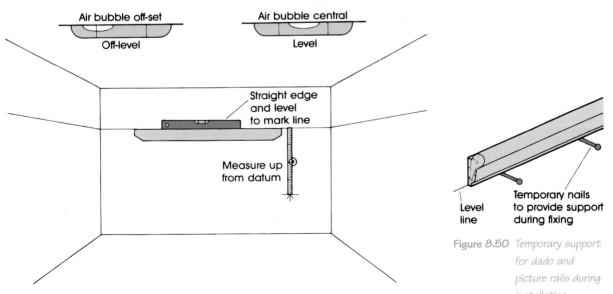

Air bubble off-set

Off-level

Air bubble central

Level

Straight edge and level to mark line

Measure up from datum

Level line

Temporary nails to provide support during fixing

Figure 8.50 Temporary support for dado and picture rails during installation

Figure 8.49 Marking positions of dado and picture rails

Simple sections can be scribed at internal corners, as are skirtings; otherwise, use mitres. External corners should be mitred.

Fixings are normally direct to the wall surface at about 400 mm centres, typically using typically either 50 mm cut nails or 50 mm masonry nails depending on the hardness of the wall. 50 mm oval or lost-heads can be used when fixing mouldings to timber studwork partitions.

Mitres around external corners should be secured with nails through their edge. Typically 38 mm ovals or lost heads are used for this purpose.

Remember, all nails should be punched below the surface ready for subsequent filling by the painter.

 Estimating materials

Determining the amount of trim required for any particular task is a fairly simple process, if the following procedures are used:

Architraves

The jambs or legs in most situations can be taken to be 2100 mm long. The head can be taken to be 1000 mm. These lengths assume a standard full-size door and include an allowance for mitring the ends. Thus the length of architrave required for one face of a door lining/frame is 5200 mm or 5.2 m.

Multiply this figure by the number of architrave sets to be fixed to determine the total metres run required, say 8 sets, both sides of four door openings:

$5.2 \times 8 = 41.6$ m

Skirtings

Skirtings and other horizontal trim can be estimated from the perimeter. This is found by adding up the lengths of the walls in the area. The widths of any doorways and other openings are taken away to give the actual metres run required.

example

Determine the total length of skirting required for the room illustrated in Figure 8.51.

> **Perimeter = 2 + 3.6 + 2.5 + 1.6 + 0.5 + 2 = 12.2 m**
>
> **Total metres run required = 12.2 − 0.8 (door opening) = 11.4 m**

An allowance of 10% for cutting and waste is normally included in any estimate for horizontal moulding.

Determine the total metres run of skirting required for the run shown including an allowance of 10% for cutting and waste.

> **Total metres run required = 11.4 m**
>
> **Total metres run required including a 10% cutting and waste allowance = 11.4 × 1.1 = 12.54 m**
>
> **say 12.5 m**

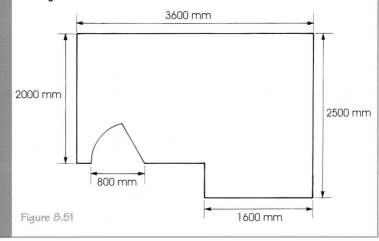

Figure 8.51

activity

Determine the total metres run of skirting required for the room shown in Figure 8.52. Include an allowance of 10% for cutting and waste.

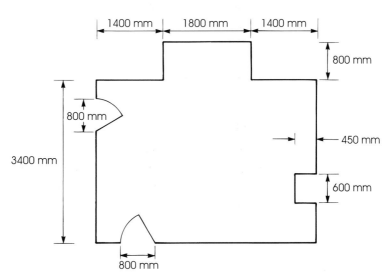

Figure 8.52

Determine the total metres run of architrave required for both faces of the doors which open into the room shown.

measuring up

1. State the reason for using architraves and skirtings.

2. Define the term 'margin' and state why it is used.

3. State why architraves are mitred and not butt jointed.

4. Sketch a plinth block and state a reason for its use.

5. Describe a situation where architraves and skirtings may be scribed.

6. Produce a sketch to show a situation where a Scotia mould may be used in place of one architrave jamb.

7. Name the preferred joint for the internal corners of skirtings, and state the reason for this preference.

8. Name the joint used for lengthening skirtings.

9. State the reason for kerfing the back of skirtings.

10. Describe a method of fixing hardwood skirting to timber grounds.

Index